The Logica Yearbook 2021

The Logica Yearbook 2021

Edited by

Igor Sedlár

© Individual authors and College Publications 2022
All rights reserved.

ISBN 978-1-84890-

College Publications
Scientific Director: Dov Gabbay
Managing Director: Jane Spurr

www.collegepublications.co.uk

Original cover design by Laraine Welch

All rights reserved. No part of this publication may be reproduced, stored in a retrieval system or transmitted in any form, or by any means, electronic, mechanical, photocopying, recording or otherwise without prior permission, in writing, from the publisher.

Preface

This volume contains peer-reviewed papers based on selected contributions presented at the conference *Logica 2021*, which took place in the Hejnice Monastery, Czech Republic, on 27 September – 1 October 2021.

The Logica conference series started in 1987. Typically taking place in a quiet venue housing all participants together, Logica aims at fostering fruitful exchanges between logicians of different specializations and generations, including students.

The programme of Logica 2021 comprised more than thirty lectures, including those given by our distinguished invited speakers: Diderik Batens, Katalin Bimbó, Rosalie Iemhoff and Christian Fermüller. A two-part tutorial lecture was given by Andrew Tedder.

On behalf of the Logica co-chair Vít Punčochář and the whole Logica organising team, I would like to thank the Institute of Philosophy of the Czech Academy of Sciences for its support of the Logica conference series, the staff of Hejnice Monastery for their hospitality and friendly assistance, and the Czech Science Foundation for financial support (grant no. 20-18675S). Vít and I are grateful to the members of the programme committee for evaluating the abstracts submitted to the conference, Katřina Krusová for administrative and practical assistance, and Ondrej Majer for designing the Logica T-shirts and other materials. As the editor of this volume, I am grateful to the reviewers of the papers for their time and valuable advice, and to College Publications and its managing director, Jane Spurr, for our pleasant cooperation during the preparation of this volume. Last but not least, I would like to thank authors of the papers included in this volume for their contributions and collaboration during the editorial process.

Prague, July 2021   Igor Sedlár

Table of Contents

Probability and Degrees of Truth 1
   Paolo Baldi and Hykel Hosni

On the Origins of Gaggle Theory 19
   Katalin Bimbó

Simple Semantics for Logics of Indeterminate Epistemic Closure 37
   Colin R. Caret

Towards Tractable Approximations to Many-Valued Logics: The Case
of First Degree Entailment 57
   Marcello D'Agostino and Alejandro Solares-Rojas

Revisiting Brandom's Incompatibility Semantics 77
   Christian G. Fermüller and Johannes Hafner

Reasoning in Commutative Kleene Algebras from *-Free Hypotheses . 99
   Stepan L. Kuznetsov

Hyperdoctrine Semantics: An Invitation 115
   Shay Logan and Graham Leach-Krouse

Truthmaker Semantics for Infectious Logics 135
   Thomas Randriamahazaka

Decidability in Proof-Theoretic Validity 153
   Will Stafford

Deontic Modal Expressivism: Proof-Theoretic and Model-Theoretic
Views 167
   Preston Stovall

Probability and Degrees of Truth

PAOLO BALDI[1] AND HYKEL HOSNI

Abstract: We argue in favour of interpreting degrees of truth in terms of objective probabilities, while retaining truth-functionality. We back our proposal with a selection of examples taken from a variety of recent approaches to quantitative uncertain reasoning.

Keywords: probability, degrees of truth, fuzzy logic

1 Introduction

Probability and intermediate degrees of truth expand the reach of logical methods to realistic reasoning agents. While they both assign values in the real unit interval to formulas of a logical language, the two concepts have very different interpretations and intended applications. The relation between them has been widely explored and clarified in the last decades (Hájek, 1998; Hisdal, 1988; Lawry, 2014; Paris, 2000).

In the AI literature (Dubois & Prade, 2001; Pearl, 1988), probabilities have been used mostly as measures of the *subjective uncertainty* of an agent, in the face of an incomplete knowledge of the facts of interest. In other sciences, there is a variety of alternative interpretations, also playing a prominent role. In particular, probabilities are used as *objective* measures, for instance in the study of random processes and in the measurement of errors (Hájek, 2019).

The interpretation of intermediate degrees of truth is similarly rich at competing interpretations. One point of consensus in the literature (Dubois, 2011; Dubois & Prade, 2001; Paris, 2000) is that intermediate degrees of truth are not adequate as measures of subjective uncertainty, and should thus be sharply distinguished from probabilities.

---

[1] Funded by the Italian Ministry of University (MUR) as part of the the PRIN 2017 project no. 20173YP4N3. The author also wishes to thank Christian Fermüller, for helping him shaping his views on the topics of vagueness and degrees of truth, in various insightful discussions during the years.

This is formally reflected by the fact that valuations of formulas admitting intermediate degrees of truth, for instance in the $[0,1]$ interval, are *truth-functional*, while probabilities are not. There is thus a strong case for *not* identifying "the truth value of a proposition p" with "the probability that p is the case".

While we agree with this point, we also believe that there is a viable interpretation of degrees of truth as probabilities, while retaining truth-functionality. In this work we make a case for such interpretation, and frame it against recent developments in the literature.

In Section 2, we recall recent results (Baldi, Cintula, & Noguera, 2020), relating logics for reasoning with probability and logics in the family of *fuzzy logics*, that is, the logics for reasoning with intermediate degrees of truth (Cintula, Fermüller, Hájek, & Noguera, 2011 – 2015). In Section 3, we put forward an interpretation of degrees of truth as measures of closeness to a threshold (or, dually, the complement of a magnitude of error), and argue that a reasonable interpretation of such measure is obtained using probabilities. In Section 4, we put our ideas to use for the modeling of vague propositions, based on the interaction of degrees of truth and probabilities. We end then with conclusions and suggestions for future work.

1.1 Related works

Various interpretations of degrees of truth based on probability have been proposed in the literature. An early example of combining a probabilistic-based interpretation of degrees of truth with assumptions of truth-functionality can be found in (Hisdal, 1988). There, one interprets degrees of truth as *conditional* probabilities, namely as the probability than an agent accepts a proposition such as "John is tall", conditional on the underlying facts, such as "John is 1.75m tall". A more recent reformulation of this idea, in the setting of coherent-based approaches to probability can also be found in (Coletti & Scozzafava, 2004),

In the philosophical literature, Edgington (1997) proposes that degrees of truth (or, in her terminology, *verities*) for vague propositions have the formal structure of probabilities, and thereby completely rejects truth-functionality. Similar complaints with truth-functionality can also be found in the linguistic literature (Sauerland, 2011).

Our work is strongly inspired by the approach put forward by Giles (1982), and revisited in (Fermüller & Metcalfe, 2009). Giles' approach assigns indeed probabilities to atomic propositions, but interprets them in a subjective sense, i.e., in terms of an agent betting on a certain kind of experiments, that serve for the assessment of the proposition under evaluation. Truth-functionality in this setting is still retained, by way of a game-theoretic interpretation of the connectives of the logical language.

2 Fuzzy Logic for reasoning about probabilities

Formal systems for combining logic and probability have been introduced since the very beginning of modern mathematical logic (Boole, 1854). Contemporary approaches (Haenni, Romeijn, Wheeler, & Williamson, 2011; Hailperin, 1996; Paris, 1994; J. Williamson, 2017) maintain the language of classical logic and seek methods to infer the probabilities, or sets of probabilities to be attached to a conclusion, given some constraints on a set of premises. This effectively leads to a generalization of classical logic.

For the purpose of this paper, we focus on a different approach, which takes probabilities as internal to the logical language. The language of classical logic is here enriched with modal formulas, standing for arithmetic operations involving probabilities. A prominent example of this approach is to be found in (Fagin, Halpern, & Megiddo, 1990) and subsequent works along that line (Halpern, 2003; Ognjanović, Rašković, & Marković, 2016).

These systems employ a *two-layered* modal syntax. The first such layer expresses classical events by means of the syntax of propositional classical logic; the second defines atomic statements as linear or polynomial inequalities of probabilities of the classical events. Each of these inequalities can be seen as the application of a multimodal operator on classical formulas. Finally, the atomic modal statements are closed again under classical connectives.

The following are examples of formulas in the language involving linear inequalities:

$$P(\varphi) \geq 0.5 \qquad (P(\varphi) + 0.3 \geq P(\psi)) \wedge (P(\varphi) \leq P(\chi) - 0.1).$$

Despite dealing with the intrinsically graded notion of probability, the semantics of these logics remains essentially bivalent: each formula can only be either true or false. Note also that $P(\varphi)$ alone does not constitute a formula of the language in this setting.

A different approach to probabilistic reasoning relies on simpler formal languages, which make use of the ideas of *mathematical fuzzy logic*. In particular in (Hájek, Godo, & Esteva, 1995), see also (Hájek, 1998), the authors introduced two-layered logics for probabilistic reasoning, based on the connectives of Łukasiewicz and product logic. The lower layer of this logic is the same as in the previous approach, namely classical logic. However, the atomic modal formulas are just sentences of the form $P(\varphi)$, i.e. "φ is probable" and the truth degree of such formula can be any value in $[0,1]$. This is clearly identified with the probability of φ. Then, the upper layer combines atomic modal formulas using suitable connectives of a fuzzy logic.

Here again a two-layered modal syntax is used, which is, however, radically simplified. Indeed, it employs only one monadic modality (for "is probable"), instead of infinitely-many polyadic modalities, and it shifts the syntactical complexity of the atomic statements to the degrees of truth in the semantics of the fuzzy logic in question.

These two modal approaches to probabilistic reasoning, the one based on classical two-layered modal logic, and that based on fuzzy modalities, are strictly connected, as shown in (Baldi, Cintula, & Noguera, 2020). We will report some results from there, but refer the interested reader to the paper for further details.

We denote by $\models_Ł$ the consequence relation of Łukasiewicz logic, and by $\models_{Ł_\triangle}$ that of Łukasiewicz logic with the addition of the $\triangle$ connective (Běhounek, Cintula, & Hájek, 2011). Furthermore, we denote by $\models_P$ the consequence relation in (Fagin et al., 1990) expressing linear inequalities involving probability, and by $\models_{FPŁ_\triangle}$ the modal fuzzy logic for reasoning with probability, where the upper layer uses the connectives of $Ł^\triangle$. The work (Baldi, Cintula, & Noguera, 2020) provides mutual translations $^\bullet, ^{\bullet\bullet}$ between the two families of modal logics of probability, in particular showing that:

1. For a set of modal formulas $\Gamma \cup \{\delta\}$ of P:

$$\Gamma \models_P \delta \text{ if and only if } \Gamma^\bullet \models_{FPŁ_\triangle} \delta^\bullet.$$

2. For a set of modal formulas $\Gamma \cup \{\delta\}$ of $FPŁ_\triangle$:

$$\Gamma \models_{FPŁ_\triangle} \delta \text{ if and only if } \Gamma^{\bullet\bullet} \models_P \delta^{\bullet\bullet}.$$

An inspection of the proof in (Baldi, Cintula, & Noguera, 2020) shows that the argument can be pushed further, and one can similarly devise translations *,**, so that even propositional Łukasiewicz logic can be used as a logic for reasoning about probability. Thus, we also have:

3. For a finite set $\Gamma \cup \{\delta\}$ of formulas of Łukasiewicz logic :

$$\Gamma \vDash_{\text{Ł}_\triangle} \delta \text{ if and only if } \Gamma^* \vDash_{\text{FPŁ}_\triangle} \delta^*.$$

4. For a finite set $\Gamma \cup \{\delta\}$ of formulas of P :

$$\Gamma \vDash_{\text{FPŁ}_\triangle} \delta \text{ if and only if } \Gamma^{**} \vDash_{\text{Ł}_\triangle} \delta^{**},$$

where Γ^{**} contains further formulas, accounting for the axioms of probability, in addition to the translation of the formulas in Γ.

The following example illustrates how the translations between fuzzy logics and classical modal logics for probability work.

Example 1 Assume we are given in P a modal probabilistic formula, such as

$$P(\varphi) \leq P(\psi).$$

Applying the translation • this will give[2] in $\text{FPŁ}_\triangle$

$$P(\varphi) \to P(\psi),$$

where the inequality becomes now the implication connective of Łukasiewicz logic.

Going further and applying the translation $**$, we obtain the formula in Łukasiewicz logic $p_\varphi \to q_\psi$ that essentially "forgets" about the lower layer formulas. The converse is conceptually interesting: In translating a propositional variable p of Łukasiewicz logic, we assume that p "hides" a probabilistic modality over some lower layer formulas. Applying the translations $*, \bullet\bullet$ we have indeed:

$$p \to q \; \mapsto^* \qquad P(\varphi) \to P(\psi) \; \mapsto^{\bullet\bullet} \qquad P(\varphi) \leq P(\psi).$$

This permits to interpret some modal fuzzy logics as logics of probabilistic reasoning, but also, in the converse direction, to see reasoning with fuzzy logic as a form of probabilistic reasoning. More precisely, fuzzy logics, such as Łukasiewicz logic, can be seen as two-layered probabilistic modal logics which have "forgotten" about their lower layer. Each propositional variable p can be thought of as a compact way of expressing a formula $P(\varphi)$,

---

[2] The translation in (Baldi, Cintula, & Noguera, 2020) also requires to prefix the formula with the connective $\triangle$, but we omit it here, since this is not essential for our purposes here.

where the formula φ of classical logic has been "forgotten", and the P is a modality standing for a measure of probability. The connectives are then ways to combine the modal formulas, i.e. to encode arithmetic computations with probabilities.

In other terms, the deductive system and semantics of Łukasiewicz logic can be used to perform truth-functional reasoning *about* probabilities. Note that this is consistent with the fact that probabilities are not truth-functional. In the next section we will follow this suggestion, and propose a way to uncover what kind of probability is hidden behind a truth degree, starting from atomic propositions in fuzzy logics.

3 Interpreting fuzzy valuations and degrees of truth

Let us fix a minimal language for a fuzzy logic, with propositional variables $p, q, \ldots$ and connectives $\odot, \oplus, \to, \sim$. Unlike what is typically done in the literature, we will reserve the symbols $\wedge, \vee, \neg$ only for the connectives acting on classical formulas. We illustrate our idea for the interpretation of degrees of truth with an example, in a series of steps.

First, let us take a propositional variable p, standing for the coarse-grained formalization of a natural language assertion such as "John is tall". In natural language, the assertion contains an adjective that is graded, and the degrees of such adjective can be taken to be the possible heights. This does not yet amount to the assumption that the proposition "John is tall" has an intermediate degree of truth.

Now, let us assume that an agent is provided a unique model for interpreting p. This means that the agent is in the *extremely idealized* position of knowing both : 1) everything there is to know about the meaning of being tall; 2) what counts as correctly asserting p.

Since p involves a graded adjective, we assume that, according to the model, one correctly asserts p when, say, "John is taller than α_k" where α_k is some threshold value (for instance, 1.80 meters). Similarly, asserting $\neg p$ might mean in this setting "John is not taller then α_1" where α_1 is another threshold value, say 1.60. Let us denote by $[p]$ John's height, i.e. the "degree of tallness", not yet to be identified with a *truth degree*. Using the two threshold values, we can then provide a first notion of model $\mathbf{M}^*$ for p, with the following 3-valued interpretation $v_{\mathbf{M}^*}$:

$$v_{\mathbf{M}^*}(p) = \begin{cases} 1 \text{ if } [p] \geq 1.80 \\ 0 \text{ if } [p] < 1.60 \\ * \text{ otherwise.} \end{cases}$$

The 3-valued interpretation is obtained at a cost: the semantic information in $[p]$ has been lost. Assume, for instance, that p stands for "John is tall" and q stands for "Mark is tall". If $[p] = 1.76$ and $[q] = 1.61$, we will have $v_{\mathbf{M}^*}(p) = v_{\mathbf{M}^*}(q) = *$, and we will thereby lose any information about the relative heights and the magnitude of such difference.

As a radical way to preserve this information, one might change both the language and the semantics, and use instead a classical language and a classical model. This can be achieved by replacing the coarse grained propositional variable p with a set of fine grained classical propositional variables, say $p_{\geq \alpha_i}$, whose intended meaning is "John is taller than α_i". We would then have a classical evaluation $v_{\mathbf{M}^c}$, such that, if John is 1.76 meters tall, then $v_{\mathbf{M}^c}(p_{\geq 1.76}) = 1$, and $v_{\mathbf{M}^c}(p_{\geq 1.77}) = 0$.

As for the case of the logic P in the previous section, the price to pay here is a much more complicated formal language, which does not allow us to express "John is tall" directly and thus takes us much farther from natural language.

The use of intermediate degrees of truth will allow us to take the best of both approaches: on the one hand, we keep a coarse grained language, as in the three-valued case, while on the other hand, we can employ the fine-grained distinction of the classical approach, hiding them in the semantics of truth values.

Let us illustrate this point, by providing a truth-degree based model $\mathbf{M}$, assigning the valuation $v_{\mathbf{M}}(p)$ to our assertion p of the form "John is tall". Note that we restrict throughout the paper to propositional logic, and, for this section, we focus only on propositional variables. We take a fuzzy valuation to provide:

- For each proposition p, a classical proposition:

$$p^{id} := \quad p_{\geq \alpha_1} \wedge \cdots \wedge p_{\geq \alpha_n}.$$

- A classical interpretation $v_{\mathbf{M}^c}$ of the conjuncts $p_{\geq \alpha_i}$.

The first item is a translation, as a classical formula, of the coarse grained assertion "John is tall". This is taken to mean that John passes all the thresholds for tallness $p_{\geq a_i}$ contemplated by the model. Now, assume that the second item, the classical interpretation $\mathbf{M}^c$, is such that

$$v_{\mathbf{M}^c}(p_{\geq 1.66}) = 1, \ldots, v_{\mathbf{M}^c}(p_{\geq 1.76}) = 1, v_{\mathbf{M}^c}(p_{\geq 1.77}) = 0, \ldots v_{\mathbf{M}^c}(p_{\geq 1.86}) = 0.$$

Under this classical evaluation, clearly $v_{\mathbf{M}_c}(p^{id}) = 0$, thus determining a loss of information, as in the example with the 3-valued case. We will then construe the intermediate truth-value $v_{\mathbf{M}}(p)$ as a compact record of how close to the truth of the classical formula p^{id} an agent is, when the classical $\mathbf{M}^c$ is actually the case.

Our proposal is thus to evaluate any propositional variable p by:

$$v_{\mathbf{M}}(p) := Cl_{\mathbf{M}^c}(p^{id}),$$

where by $Cl_{\mathbf{M}^c}(p^{id})$ we denote the function measuring how close is $\mathbf{M}^c$ to the truth of p^{id}. Now, clearly various options are left open to define the function $Cl_{\mathbf{M}^c}$, and consequently, the fuzzy evaluation $v_{\mathbf{M}}$ over p.

An obvious way of defining it, is just by counting the ratio of conjuncts in p^{id} which are true according to the classical interpretation $\mathbf{M}^c$. This results in the following:

$$Cl_{\mathbf{M}^c}(p^{id}) = \frac{|\{p_{\alpha_i} \mid v_{\mathbf{M}^c}(p_{\alpha_i}) = 1\}|}{|\{p_{\alpha_i} \mid p_{\alpha_i} \in p^{id}\}|} = \frac{k}{n},$$

where, with an abuse of notation, by $p_{\alpha_i} \in p^{id}$, we mean that p_{α_i} is one of the conjuncts of p^{id}. The above is a probability in the classical sense, that assumes a uniform distribution over the $p_{\geq \alpha_i}$, and just counts the number of favourable cases over the possible ones.

While this is a probability, we are still assuming that a fuzzy model $\mathbf{M}$ is given, which specify all there is to know about p. There is thus no uncertainty on the side of the agent, neither about whether p occurs, nor about which of the $p_{\geq \alpha_i}$ is a correct classical interpretation for p. As we will see in Section 4, this latter interpretation is more appropriate when dealing with vagueness, and is indeed at the core of our rendering of approaches based on supervaluations (Fermüller & Kosik, 2006; Fine, 1975) and voting semantics (Lawry, 1998).

The truth value $v_{\mathbf{M}}(p)$ is thus neither the probability of p, nor of $p_{\geq \alpha_i}$, nor of p^{id}. Formally speaking, it is the probability that, by picking randomly a conjunct $p_{\geq \alpha_i}$ in p^{id}, one asserts a (classically) true proposition, according to the evaluation $\mathbf{M}^c$. We can thus read the truth degree as the probability that, even though an agent asserts an incorrect formula (p^{id} is classically false), this will not be found out after a random check. Since the probability used as a truth degree involves no subjective uncertainty, and it is rather used as a device to measure how close $\mathbf{M}^c$ is to making p^{id} true, we can take it as form of objective[3] probability.

Note that there are other options available for defining $Cl_{\mathbf{M}^c}$. For instance, one might impose more stringent standards for closeness, and require more than one sampling (with replacement) among the conjuncts in p^{id}. Similar ideas have been developed for the modeling of fuzzy quantifiers in (Baldi, Fermüller, & Hofer, 2020; Fermüller & Roschger, 2014) and can be taken as sources of inspiration for addressing this issue.

Remark 1 The probabilistic interpretation put forward here paves the way for using "degrees of truth", even in scenarios which do not contain graded propositions. Indeed, any conjunction in classical propositional logic can be held in this sense to be more or less false, depending on how many of the conjuncts hold. The same holds for a universally quantified sentence in first-order classical logic, say $\forall x A(x)$ where we might want to distinguish, e.g. between: a) a scenario where $A(x)$ holds for all but one element of a domain and b) a scenario in which $A(x)$ does not hold for any element of the domain. While the classical truth value assigned in both cases is 0, the needed information to distinguish a) from b) is instead provided by the (objective) probability that a randomly picked element of the domain satisfies the predicate. Let us note that Adams (1998) calls indeed "degree of truth" such a probability, to be sharply distinguished from the subjective probability that an agent might assign to the sentence $\forall x A(x)$.

To summarize, we argued that degrees of truth can be interpreted as objective probabilities, and that they allow to keep track of fine-grained information concerning closeness to truth. Furthermore, under our reading, the choice between a fuzzy, a classical or a three-valued evaluation is ultimately to be grounded on the kind of information we want to record in the semantics, rather than on the nature of the natural language expressions of concern.

---

[3] In our use, the expression "objective probability" does not point to a frequentist or a propensity interpretation of probability, as is typically the case in the literature. Our use is in some respect more closely connected to logical intepretations of probability (Hájek, 2019).

3.1 Truth-functionality

So far we have only discussed a probabilistic reading of the degree of truth that a fuzzy valuation $v_\mathbf{M}(p)$ assign to a propositional variable p. An essential property of fuzzy valuations is that they should behave truth-functionally, i.e. for each formula $\star(\varphi_1,\ldots,\varphi_n)$, with n-ary connective $\star$, the evaluation is to be obtained as a fixed function of the evaluations of its subformulas φ_i. This contrasts with probabilities, which are generally non-truth functional, thus seeming to contradict our probabilistic reading of intermediate truth values.

This apparent contradiction is actually solved, if we take fuzzy formulas as formulas for reasoning *about* probabilities, as for the logics in Section 2. We may just impose truth-functionality, by letting $v_\mathbf{M}(p_i) = Cl_{\mathbf{M}^c}(p_i^{id})$ for any propositional variable p_i, and then, for any connective $\star$, $v_\mathbf{M}(\star(\varphi_1,\ldots,\varphi_n)) = f^\star(v_\mathbf{M}(\varphi_1),\ldots,v_\mathbf{M}(\varphi_n))$, where $f^\star$ is a suitable truth-function interpreting the connective $\star$.

This means that we are safely restricting the definition of the classical formula p^{id} and the measure of closeness $cl_{\mathbf{M}^c}$ only to propositional variables. We do not need to define, say, a classical counterpart for a complex formula such as $(p \wedge q)^{id}$, $(p \wedge \neg p)^{id}$ or $(p \wedge q)^{id}$. Relying on that kind of construction would indeed be problematic, since it would easily lead to conflicts with the assumption of truth-functionality (Dubois, 2011; Smith, 2015).

This suffices for *enforcing* truth-functionality in our approach. We might however, obtain it in a more principled way in our framework.

Recall that, for us, a fuzzy model $\mathbf{M}$ can be read as assigning to a propositional variable p a classical formula, say $(p^{id}) := p_{\geq \alpha_1} \wedge \cdots \wedge p_{\geq \alpha_m}$ and a classical valuation $v_{\mathbf{M}^c}$. Assume to have $v_{\mathbf{M}^c}(p_{\geq \alpha_i}) = 1$ for each $i \leq k < m$, and $v_{\mathbf{M}^c}(p_{\geq \alpha_i}) = 0$ otherwise. The valuation of p results from the probability of picking a true p_{α_i} by random sampling, i.e. in this case k/m.

We extend this reading in a truth-functional way, by just imposing that an *independent sampling* is performed for each *occurrence* of a propositional variable such as p, in a fuzzy formula. For instance, given a formula such as $p \odot p$, we take $v_\mathbf{M}(p \odot p)$ as the probability of picking two true conjuncts $p_{\geq \alpha_i}$, when sampling twice the conjuncts in p^{id}. This leads then to $v_\mathbf{M}(p \odot p) = v_\mathbf{M}(p) \cdot v_\mathbf{M}(p)$.

Note that, if assessing the value of p was a matter of uncertainty, and sampling was a method to resolve the uncertainty, one sampling would have sufficed, to settle whether p occurred or not. This is a crucial distinction between probabilities over classical logic and degrees of truth.

Probability and Degrees of Truth

In our setting, instead, we assume that an agent knows all there is to know about p, but wants to keep track of how close to the truth she is when asserting p^{id} when $\mathbf{M}^c$ is the case. $p \odot p$ stands for asserting p^{id} twice, hence her closeness to truth (complement of a magnitude of error) will need to be taken into account twice.

Similarly, take two fuzzy propositions p, q, with the corresponding :

$$p^{id} := \{p_{\geq \alpha_1}, \ldots, p_{\geq \alpha_n}\} \text{ and } q^{id} := \{q_{\geq \alpha_1}, \ldots, q_{\geq \alpha_m}\},$$

and classical evaluation $\mathbf{M}^c$. We may let then :

$$v_{\mathbf{M}}(p \odot q) = \frac{|\{(p_{\geq \alpha_i}, q_{\geq \beta_j}) \mid v_{\mathbf{M}^c}(p_{\geq \alpha_i} \wedge q_{\geq \beta_j}) = 1\}|}{|\{(p_{\geq \alpha_i}, q_{\geq \beta_j}) \mid p_{\geq \alpha_i} \in p^{id}, q_{\geq \beta_j} \in q^{id}\}|}$$

i.e. we consider *independently* randomly picking a $p_{\geq \alpha_i}$ among the conjunct in p^{id} and a $q_{\geq \beta_j}$ among the conjuncts in q^{id}. The truth value of the conjunction will thus record the fact that both randomly picked conjuncts are (classically) true in $\mathbf{M}^c$. This justifies letting:

$$v_{\mathbf{M}}(p \odot q) = Cl_{\mathbf{M}^c}(p^{id}) \cdot Cl_{\mathbf{M}^c}(q^{id}) = v_{\mathbf{M}}(p) \cdot v_{\mathbf{M}}(q).$$

More generally, assume to have any formula $\varphi(p_1, \ldots, p_m)$, which is built via the connectives $\odot, \oplus, \sim$, where $p_1, \ldots, p_m$ are all the *occurrences* of propositional variables, each with corresponding formula p_i^{id}. Let us denote by φ^c the classical formulas obtained by replacing each connective $\odot, \oplus, \sim$ by the classical $\wedge, \vee, \neg$. We will then let:

$$v_{\mathbf{M}}(\varphi) = \frac{|\{(p_{1 \geq \alpha_{i_1}}, \ldots, p_{m \geq \alpha_{i_m}}) \mid v_{\mathbf{M}^c}(\varphi^c(p_{1 \geq \alpha_{i_1}}, \ldots, p_{m \geq \alpha_{i_m}})) = 1\}|}{|\{(p_{1 \geq \alpha_{i_1}}, \ldots, p_{m \geq \alpha_{i_m}}) \mid p_{1 \geq \alpha_{i_1}} \in p_1^{id}, \ldots, p_{m \geq \alpha_{i_m}} \in p_m^{id}\}|},$$

where each $i_1, \ldots i_m$ is an index that picks any of the conjuncts in the corresponding formula $p_1^{id}, \ldots, p_m^{id}$.

This principle leads to interpret a disjunction as $v(\varphi \oplus \psi) = v(\varphi) + v(\psi) - v(\varphi) \cdot v(\psi)$, and negation as $v(\sim \varphi) = 1 - v(\varphi)$. Note that, while the interpretation of conjunction and disjunction in our approach coincide with that of *product logic*, the function for negation is instead that of Łukasiewicz logic, see (Běhounek et al., 2011).

Different sets of motivations can also be given in our framework for using the conjunction of Łukasiewicz, e.g. by considering ideas developed for Giles' game, and the related interpretation of hypersequents in Łukasiewicz logic (Fermüller & Metcalfe, 2009).

The ideas sketched above can also be included in the formal language of a fuzzy logic, by suitably enriching it. The resulting formal language will have some analogies with those in Section 2, but it does not need the explicit addition of modalities which correspond to the function $Cl_{\mathbf{M}^c}$.

We consider an extension of a fuzzy logic, such as product or Łukasiewicz logic, with an additional non-associative connective π (Fermüller, 2014) (which is actually a distinct n-ary connective π_n for any $n \in \mathbb{N}$) to be interpreted as an average of (classical) truth values, i.e.:

$$v_{\mathbf{M}}(\pi(p_1,\ldots,p_n)) = \sum_{i=1}^{n} \frac{v_{\mathbf{M}^c}(p_i)}{n}.$$

Note that in this case, we take the p_is to be classical propositional variables and $\mathbf{M}^c$ to be the classical evaluation associated to $\mathbf{M}$. Now, for any propositional variable p in a fuzzy logic, with $p^{id} := p_1 \wedge \cdots \wedge p_n$, we replace p with $\pi(p_1,\ldots,p_n)$. We will have then:

$$v_{\mathbf{M}}(p) = v_{\mathbf{M}}(Cl_{\mathbf{M}^c}(p^{id})) = v_{\mathbf{M}}(Cl_{\mathbf{M}^c}(p_1 \wedge \cdots \wedge p_n)) = v_{\mathbf{M}}(\pi(p_1,\ldots,p_n)).$$

To be more precise, we can devise a formal language as follows:

$$\Phi ::= \pi(p,\ldots,p) \mid \Phi \odot \Phi \mid \Phi \oplus \Phi \mid \Phi \to \Phi \mid \sim \Phi$$

where propositional variables are only meant to be classical; atomic formulas of fuzzy logic are obtained via π applied to propositional variables, and finally complex fuzzy logical formulas are obtained combining the π-prefixed formulas via the connectives of a fuzzy logic.

The language is thus reminiscent of the two-layered language in Section 2, with the lower layer made of propositional variables, and the atomic formulas of the upper layer of the form $\pi(p,\ldots,p)$.

Note that the evaluation of a language as the above needs only a classical valuation $v_{\mathbf{M}^c}$ of the propositional variables. Indeed, the conjuncts of the formula p^{id} are included in the language, and the intermediate truth degrees are then completely determined, by using the connective π and then the truth-functions for the corresponding connectives.

4 Vagueness

So far, we have treated a fuzzy model as having sharp thresholds for accepted, rejected and undecided propositions. Following (Cintula, Noguera, &

Smith, 2017), we have thus seen fuzzy logics as dealing mainly with *graded* propositions rather than *vague* ones.

We believe that there is, however, a tight, albeit indirect, link with vagueness. When propositions are vague, we have stronger reasons to keep track of errors (or closeness to thresholds), since any threshold which has been fixed for the acceptance and rejection of a proposition is purely conventional and idealized.

Recall that, in our framework, a fuzzy model associates to each classical proposition p a formula p^{id}, for encoding the thresholds. Then it evaluates the closeness of a (classical) valuation $\mathbf{M}^c$ to the truth of p^{id}.

We take vagueness to be primarily a form of semantic indeterminacy (Dubois, 2011; Lawry, 2014), where an agent does not know the right thresholds to be applied. We render this via a set of fuzzy models, expressing the agent's subjective uncertainty, perhaps unresolvable in principle. Since the uncertainty is only of a semantic nature, each such model will determine the same classical $\mathbf{M}^c$, but a different p^{id}, for any proposition p. We denote in the following by p^{id_j} the ideal propositions corresponding to a fuzzy model indexed by j. Following the terminology in (Fine, 1975) we will call each such p^{id_j} a *precisification*.

This background justifies the use of a probability, this time a purely subjective one, measuring which of the fuzzy model (i.e which p^{id_j}) is more likely to be appropriate for the agent. Once one of the p^{id_j} is chosen for a propositional variable p, one can again evaluate the truth value, by way of an objective probability, as it was done in the previous section.

This is not to say that we need to postulate the existence a "correct" threshold. We assume nevertheless that an agent operates as if there were such an ideal threshold in a particular context, of which she does not know the exact location. Thus we follow what is called an *epistemic* stance towards vagueness in (Lawry, 2014)[4].

The two steps approach to vagueness sketched above is also reminiscent of the ideas of fuzzy plurivaluationism (Smith, 2008), where vague propositions are interpreted via a *plurality* of *fuzzy* models. These are aimed at addressing what Smith dubs the "location" and "jolt" problems. In our framework, the location problem is modelled by the set of possible thresholds, quantified by a subjective probability, and in a non truth-functional

---

[4]This is to be distinguished from Williamson's epistemicism (T. Williamson & Simons, 1992). It is not assumed that there is an unknown fact about the correct threshold, but we are nevertheless modeling an agent acting *as if* there were such unknown correct threshold, thereby using a subjective probability.

way. The jolt problem is addressed instead by measuring closeness to the selected threshold, hence truth-functionally, and possibly using an objective probability, as discussed in Section 3.

Let us illustrate our ideas on vagueness with an example. Consider again the proposition "John is tall", that we denote by the propositional variable p. This time we will take a set of fuzzy models, all with the same classical evaluation $\mathbf{M}^c$, but with different precisification formulas p^{id_j} for $j = 1, \ldots m$, providing the possible different thresholds.

We further assume that an agent has a probability distribution P over the precisifications, reflecting her belief that p^{id_j} is the correct threshold for the propositional variable p. To avoid confusion with the non-vague case of the previous section, following (Fermüller & Kosik, 2006) let us put a symbol ∘ in front of p, to indicate that we are taking an expected evaluation of p over a set of precisifications.

For any propositional variable p, with precisifications p^{id_j} for $j = 1, \ldots m$, each with probability $P(p^{id_j})$, we will then let:

$$v_{\mathbf{M}}(\circ p) = \sum_{j=1}^{m} Cl_{\mathbf{M}^c}(p^{id_j})P(p^{id_j}).$$

More generally, take any formula $\varphi = f(p_1, \ldots, p_n)$ such that, for each propositional variable p_i the precisifications are of the form $p^{id_j}_i$. Let us further assume for simplicity that $j = 1, \ldots, m$ for each i. We let then:

$$v_{\mathbf{M}}(\circ \varphi) = \sum_{i=1}^{n} \sum_{j_i=1}^{m} f(Cl_{\mathbf{M}^c}(p_1^{id_{j_1}}), \ldots, Cl_{\mathbf{M}^c}(p_n^{id_{j_n}}))P(p_1^{id_{j_1}} \wedge \cdots \wedge p_n^{id_{j_n}}),$$

where $P(p_1^{id_{j_1}} \wedge \cdots \wedge p_n^{id_{j_n}})$ is the subjective probability of picking those formulas as the correct thresholds, and, as such, need not be truth-functional.

5 Conclusions and future works

We have explored connections between degrees of truth and probabilities, and in particular between fuzzy logics and logical system for probabilistic reasoning. In particular, we have put forward a reading of degrees of truth as probabilities, following a general idea of "closeness to a threshold".

Under our reading, intermediate truth degrees are useful for keeping track of errors, which are due to the interaction of a coarse and a fine-grained level of analysis of propositions, as also stressed in (Vetterlein,

2011). Using truth-degrees allows us indeed to use a coarse-grained simple language (the variables p in the syntax of the fuzzy logic), while recording into the semantics of truth-values the deep, underlying, fine-grained classical distinctions (the conjuncts of p^{id} associated with each p).

Concerning future work, we believe that the idea of taking truth degrees as closeness to a threshold naturally lend itself beyond the probabilistic interpretation. In particular, we may drop the requirement that the closeness, and hence the truth degree, should have a minimum and a maximum, as it is the case for the $[0, 1]$ interval. We might instead have infinitely many positive and negative degrees of truth. A natural setting for these ideas will be either Abelian logic (Meyer & Slaney, 1989) or logics such as Casari's comparative logic (Paoli, 2003).

Furthermore, we plan to explore the implications of our probabilistic reading of degrees of truth for the study of generalized quantifiers (Baldi, Fermüller, & Hofer, 2020), and for related issues concerning statistical reasoning (Adams, 1998; Kyburg & Teng, 2001).

References

Adams, E. W. (1998). *A Primer of Probability Logic*. Stanford: CSLI Publications.

Baldi, P., Cintula, P., & Noguera, C. (2020). Classical and fuzzy two-layered modal logics for uncertainty: Translations and proof-theory. *International Journal of Computational Intelligence Systems*, *13*(1), 988–1001.

Baldi, P., Fermüller, C. G., & Hofer, M. F. J. (2020). On fuzzification mechanisms for unary quantification. *Fuzzy Sets and Systems*, *388*, 90–123.

Běhounek, L., Cintula, P., & Hájek. (2011). Introduction to mathematical fuzzy logic. In P. Cintula, C. Fermüller, P. Hájek, & C. Noguera (Eds.), *Handbook of Mathematical Fuzzy Logic – Volume 1* (Vol. 37, pp. 1–101). London: College Publications.

Boole, G. (1854). *An Investigation of the Laws of Thought on Which Are Founded the Mathematical Theories of Logic and Probabilities*. London: Walton and Maberly.

Cintula, P., Fermüller, C., Hájek, P., & Noguera, C. (Eds.). (2011 – 2015). *Handbook of Mathematical Fuzzy Logic* (Vol. 37-38-58). London: College Publications.

Cintula, P., Noguera, C., & Smith, N. J. J. (2017). A logical framework for graded predicates. In A. Baltag, J. Seligman, & T. Yamada (Eds.), *Logic, Rationality, and Interaction - 6th International Workshop (LORI)* (Vol. 10455, pp. 3–16). Berlin, Heidelberg: Springer.

Coletti, G., & Scozzafava, R. (2004). Conditional probability, fuzzy sets, and possibility: A unifying view. *Fuzzy Sets and Systems, 144*(1), 227–249.

Dubois, D. (2011). Have fuzzy sets anything to do with vagueness? In P. Cintula, C. G. Fermüller, L. Godo, & P. Hájek (Eds.), *Understanding Vagueness: Logical, Philosophical and Linguistic Perspectives* (pp. 311–333). London: College Publications.

Dubois, D., & Prade, H. (2001). Possibility theory, probability theory and multiple-valued logics: A clarification. *Annals of Mathematics and Artificial Intelligence, 32*, 35–66.

Edgington, D. (1997). Vagueness by degrees. In R. Keefe & P. Smith (Eds.), *Vagueness: A Reader.* Cambridge, MA: MIT Press.

Fagin, R., Halpern, J. Y., & Megiddo, N. (1990). A logic for reasoning about probabilities. *Information and Computation, 87*(1–2), 78–128.

Fermüller, C. G. (2014). Hintikka-style semantic games for fuzzy logics. In C. Beierle & C. Meghini (Eds.), *Foundations of Information and Knowledge Systems* (pp. 193–210). Cham: Springer International Publishing.

Fermüller, C. G., & Kosik, R. (2006). Combining supervaluation and degree based reasoning under vagueness. In M. Hermann & A. Voronkov (Eds.), *Logic for Programming, Artificial Intelligence, and Reasoning. 13th International Conference, LPAR 2006.* (Vol. 4246, pp. 212–226). Berlin, Heidelberg.

Fermüller, C. G., & Metcalfe, G. (2009). Giles game and proof theory for Lukasiewicz logic. *Studia Logica, 92*(1), 27–61.

Fermüller, C. G., & Roschger, C. (2014). Randomized game semantics for semi-fuzzy quantifiers. *Logic Journal of the IGPL, 22*(3), 413–439.

Fine, K. (1975). Vagueness, truth and logic. *Synthese, 30*(3/4), 265–300.

Giles, R. (1982). Semantics for fuzzy reasoning. *International Journal of Man-Machine Studies, 17*(4), 401–415.

Haenni, R., Romeijn, J.-W., Wheeler, G., & Williamson, J. (2011). *Probabilistic Logics and Probabilistic Networks* (No. 350). Dordrecht: Springer.

Hailperin, T. (1996). *Sentential Probability Logic*. Bethlehem, PA: Lehigh University Press.

Hájek, P. (1998). *Metamathematics of Fuzzy Logic*. Dordrecht: Kluwer.

Hájek, P., Godo, L., & Esteva, F. (1995). Fuzzy logic and probability. In P. Besnard & S. Hanks (Eds.), *UAI'95: Proceedings of the Eleventh Conference on Uncertainty in Artificial Intelligence* (pp. 237–244). San Francisco: Morgan Kaufmann.

Halpern, J. Y. (2003). *Reasoning About Uncertainty*. Cambridge, MA: MIT Press.

Hisdal, E. (1988). Are grades of membership probabilities? *Fuzzy Sets and Systems*, 25(3), 325—348.

Hájek, A. (2019). Interpretations of probability. In E. N. Zalta (Ed.), *The Stanford Encyclopedia of Philosophy* (Fall 2019 ed.). Metaphysics Research Lab, Stanford University. `plato.stanford.edu/entries/probability-interpret`.

Kyburg, H. E., Jr, & Teng, C. M. (2001). *Uncertain Inference*. Cambridge: Cambridge University Press.

Lawry, J. (1998). A voting mechanism for fuzzy logic. *International Journal of Approximate Reasoning*, 19(3-4), 315–333.

Lawry, J. (2014). Probability, fuzziness and borderline cases. *International Journal of Approximate Reasoning*, 55(5), 1164–1184.

Meyer, R. K., & Slaney, J. K. (1989). Abelian logic (from A to Z). In G. Priest, R. Routley, & J. Norman (Eds.), *Paraconsistent Logic. Essays on the Inconsistent.* (pp. 245–305). Munich, Hamden, Vienna: Philosophia.

Ognjanović, Z., Rašković, M., & Marković, Z. (2016). *Probability Logics*. Cham: Springer International Publishing.

Paoli, F. (2003). A really fuzzy approach to the Sorites paradox. *Synthese*, 134(3), 363–387.

Paris, J. B. (1994). *The Uncertain Reasoner's Companion: A Mathematical Perspective*. Cambridge: Cambridge University Press.

Paris, J. B. (2000). Semantics for fuzzy logic supporting truth functionality. In V. Novák & I. Perfilieva (Eds.), *Discovering the World with Fuzzy Logic* (pp. 82–104). Heidelberg: Physica.

Pearl, J. (1988). *Probabilistic Reasoning in Intelligent Systems: Networks of Plausible Inference*. San Francisco: Morgan Kaufmann.

Sauerland, U. (2011). Vagueness in language: The case against fuzzy logic revisited. In P. Cintula, C. G. Fermüller, L. Godo, & P. Hájek (Eds.), *Understanding Vagueness: Logical, Philosophical and Linguistic Perspectives* (pp. 185–198). London: College Publications.

Smith, N. J. J. (2008). *Vagueness and Degrees of Truth*. Oxford: Oxford University Press.

Smith, N. J. J. (2015). Fuzzy logics in theories of vagueness. In P. Cintula, C. Fermüller, P. Hájek, & C. Noguera (Eds.), *Handbook of Mathematical Fuzzy Logic – Volume 3* (pp. 1237–1281). London: College Publications.

Vetterlein, T. (2011). Vagueness: A mathematician's perspective. In P. Cintula, C. G. Fermüller, L. Godo, & P. Hájek (Eds.), *Understanding Vagueness: Logical, Philosophical and Linguistic Perspectives* (pp. 67–86). London: College Publications.

Williamson, J. (2017). *Lectures on Inductive Logic*. Oxford: Oxford University Press.

Williamson, T., & Simons, P. (1992). Vagueness and ignorance. *Proceedings of the Aristotelian Society, Supplementary Volumes*, 66, 145–177.

Paolo Baldi
University of Milan, Department of Philosophy
Italy
E-mail: `paolo.baldi@unimi.it`

Hykel Hosni
University of Milan, Department of Philosophy
Italy
E-mail: `hykel.hosni@unimi.it`

On the Origins of Gaggle Theory

KATALIN BIMBÓ[1]

Abstract: The *generalized Galois logic* approach (i.e., *gaggle theory*), introduced by Dunn, provides a systematic way to define semantics for many substructural logics in the form of a relational representation of their Lindenbaum algebras. We provide an overview of some conceptual antecedents that we think that likely contributed to the creation of gaggle theory. In our reconstruction, we rely on Dunn's publications and some materials deposited in the Archives of Indiana University, Bloomington, IN, U.S.A.

Keywords: intuitionistic logic, Meyer–Routley semantics, modal logic, possible world semantics, relevance logic, residuation, **R**-mingle, tense logic

1 Introduction

Dunn published a series of papers in the 1990s, in which he presented *gaggle theory*. The name "gaggle" intends to serve as a convenient pronunciation of the acronym "gGl," which abbreviates *generalized Galois logic*. Gaggle theory is, perhaps, better viewed as an *approach* to substructural and intensional logics rather than a motley collection of definitions and theorems. As a first approximation, gaggle theory aims to bring under a common theoretical umbrella the ways in which concrete set-theoretical semantics are defined for a range of logics that exceed 2-valued logic (**TV**) in some way. A slightly more precise description would mention two steps in this process. First, the Lindenbaum algebra of a logic is formed; second, that algebra is represented using a relational structure in the sense of algebraic representation theory. Moreover, gaggle theory does not simply amount to an aggregation of set-theoretic constructions for various logics. It *generalizes* existing semantics and furnishes new semantics for logics in *a systematic way* based on the algebraic properties of the logics.

---

[1] I am grateful to Vít Punčochář and Igor Sedlár, the organizers of *Logica 2021*, for asking me to give an invited talk, the content of which overlaps that of this paper. I would like to thank the audience at *Logica 2021* and an anonymous referee for their questions and comments. The research reported in this paper is partially funded by an *Insight Grant* (#435–2019–0331) awarded by the *Social Sciences and Humanities Research Council* of Canada.

Katalin Bimbó

This paper traces the *emergence of gaggle theory* in Dunn's work to the late 1970s, and points at some set-theoretical semantics that probably contributed to the formulation of the theory. The semantics we mention are—in the order of their appearance—Kripke-style semantics for modal and tense logics, **BAO**'s, Kripke's semantics for intuitionistic logic, Dunn's semantics for **R**-mingle (**RM**) and the Meyer–Routley semantics for relevance logics.

2 Some semantics as motivations for gaggle theory

2.1 Semantics for some normal modal logics

Kripke's semantics for some *normal modal logics* was first described in Kripke (1959), and then in Kripke (1963). This set-theoretical semantics is widely known now, and it was surely known in the 1960s by Dunn whose Ph.D. thesis supervisor was Nuel D. Belnap. Alan R. Anderson, Belnap and Kripke corresponded in the late 1950s. Indeed, Belnap pointed out to Kripke the decidability problem of $\mathbf{E}_\to$ in a letter dated May 31st, 1959, which Kripke solved within a few months.[2]

Let us consider the modal logic **S4** to illustrate an idea and a puzzlement.[3] The language of **S4** contains a denumerable sequence of sentence letters $\langle p_i \rangle_{i \in \omega}$. Formulas are generated by $\neg$ (negation), $\supset$ (conditional) and $\Box$ (necessity) as usual, and $\mathcal{A}, \mathcal{B}, \mathcal{C}, \ldots$ range over formulas. An axiomatic system for **S4** may be defined by adding to an axiomatization of 2-valued logic (with detachment as a rule) the following axioms and rule

(K) $\Box(\mathcal{A} \supset \mathcal{B}) \supset (\Box \mathcal{A} \supset \Box \mathcal{B})$ (T) $\Box \mathcal{A} \supset \mathcal{A}$ (4) $\Box \mathcal{A} \supset \Box \Box \mathcal{A}$
(nec) $\vdash \mathcal{A}$ implies $\vdash \Box \mathcal{A}$

The notions of a *proof* and a *theorem* are defined as usual. We limit our considerations to this simple notion of consequence in this example.

A possible world semantics for **S4** is based on a *structure*, which is a pre-ordered (or possibly, weakly partially ordered) non-empty set of worlds, $\mathfrak{F} = \langle W, R \rangle$. ($R \subseteq W \times W$, and it is a reflexive and transitive (and possibly, anti-symmetric) relation.) A *model* $\mathfrak{M}$ adds a valuation v that assigns a set of worlds to each p, that is, $v(p) \subseteq W$. The meaning of $w \in v(p)$ is that p

---

[2] Belnap's letter is preserved within Kripke's correspondence in the Kripke Archives; cf. Bimbó (2020).

[3] We harmonize the notation in our presentation of ideas from several semantics, somewhat along the lines of Bimbó and Dunn (2008). Accordingly, we might not follow the notation or the terminology of the publications we refer to.

is true in the world w. The interpretation of all the formulas is given by an extension of v, which we denote by $[\![\ ]\!]$ (omitting decorations to indicate the model, which is fixed by the context). To start with, $[\![p]\!] = \{\, w \colon w \in v(p)\,\}$.

1. $[\![\neg \mathcal{A}]\!] = W \setminus [\![\mathcal{A}]\!]$ 2. $[\![\mathcal{A} \supset \mathcal{B}]\!] = (W \setminus [\![\mathcal{A}]\!]) \cup [\![\mathcal{B}]\!]$
3. $[\![\Box \mathcal{A}]\!] = \{\, w \colon \forall w'(Rww' \Rightarrow w' \in [\![\mathcal{A}]\!])\,\}$

The notions of $\mathcal{A}$ being *true at* w (i.e., $w \in [\![\mathcal{A}]\!]$), $\mathcal{A}$ being *valid* in $\mathfrak{M}$ ($\forall w\ w \in [\![\mathcal{A}]\!]$), and $\mathcal{A}$ being *valid in a class of models* $\mathbb{C}$ ($\forall \mathfrak{M} \in \mathbb{C}$, $\mathcal{A}$ is valid in $\mathfrak{M}$) are defined as usual. We may write $\vDash_{\mathfrak{F}} \mathcal{A}$, as customary, to indicate validity in $\mathbb{C}$, where $\mathbb{C}$ is the class of models on $\mathfrak{F}$.

Theorem 1 *For any formula* $\mathcal{A}$, $\vdash_{S4} \mathcal{A}$ *iff* $\vDash_{\mathfrak{F}} \mathcal{A}$, *where* $\mathfrak{F}$ *is as above.*

The proof of this soundness and completeness theorem is fairly routine, and there are multiple published versions of it. Instead of spelling out the details, we note that the definition of $\Diamond$ (possibility) is straightforward. $\neg \Box \neg \mathcal{A}$ expresses that $\mathcal{A}$ is possible, because it is not that not-$\mathcal{A}$ is necessary. Having worked through the definition of the truth of $\neg \Box \neg \mathcal{A}$ step by step, we obtain the following clause.

4. $[\![\Diamond \mathcal{A}]\!] = \{\, w \colon \exists w'(Rww' \wedge w' \in [\![\mathcal{A}]\!])\,\}$

This means that having both $\Box$ and $\Diamond$ when they are *definable* via $\neg$ is unproblematic in the semantics of normal modal logic—in the sense that a *single accessibility relation* is sufficient to model these connectives. We may note that 3 and 4 are neatly in line with Leibniz's readings of modalities through $\forall$ and $\exists$, although he seems to have assumed that all worlds are accessible from any world (cf. Look (2016)).

Tense logics, more precisely some versions of them, introduce modality-like operators for reasoning about the *past* and the *future*. A minimal tense logic, denoted here by $\mathbf{K}_t$, has two $\Diamond$-type connectives, namely, P (sometime in the past) and F (sometime in the future). Their $\Box$-like duals are H (always in the past) and G (always in the future). The tense axioms are two instances of (K) and two instances of (4) with H and G, respectively, and a pair of axioms (B↓) and (B↑) that tie the future and the past together, so to speak. These axioms are analogs of the axiom (B) $\mathcal{A} \supset \Box \Diamond \mathcal{A}$, which is a theorem of the normal modal logics **B** and **S5**.

(B↓) PG$\mathcal{A} \supset \mathcal{A}$ (B↑) FH$\mathcal{A} \supset \mathcal{A}$

Finally, the (nec) rule is stipulated for both G and H. The (4) axioms will force the transitivity of the accessibility relation moving into the future and moving into the past. However, what is interesting from our point of view is that (B↓) and (B↑) create a very close relationship between R_G and R_H.

In particular, $R_{\mathsf{G}} = R_{\mathsf{H}}^{\smile}$, that is, the accessibility relations are each other's *converses*. Then, we may omit the subscripts and state the satisfiability conditions for G and H—reusing a previous clause for □—as follows.

5. $[\![\mathsf{G}\mathcal{A}]\!] = \{\, w \colon \forall w'(Rww' \Rightarrow w' \in [\![\mathcal{A}]\!])\,\}$
6. $[\![\mathsf{H}\mathcal{A}]\!] = \{\, w \colon \forall w'(Rw'w \Rightarrow w' \in [\![\mathcal{A}]\!])\,\}$

Rww' means informally that w' is in the future of w. If we swap the arguments as in $Rw'w$, then w' is in the past of w.[4] From the point of view of gaggle theory, it is interesting that G and H are *not definable* from each other in the context of $\mathbf{K}_t$, yet they are modeled from the same accessibility relation. However, F and H as well as P and G are each other's *residuals*. For example, for P and G, this means that $\mathsf{P}\mathcal{A} \supset \mathcal{B}$ is a theorem of $\mathbf{K}_t$ iff $\mathcal{A} \supset \mathsf{G}\mathcal{B}$ is a theorem.

2.2 Boolean algebras with operators

Boolean algebras are important beyond their basic role as the class of algebras into which the Lindenbaum algebra of **TV** falls. For instance, elementary probability theory adds a countably additive normal function on the event space and relation algebras add further operations such as relational composition and converse. Jónsson and Tarski (1951–52) introduced a lesser representation for relation algebras as part of a general representation theory for **BAO**'s. The preferred representation of a relation algebra is by binary relations, whereas their focus is on representing the operations on binary relations (the elements of a relation algebra) by relations of appropriate (i.e., one larger) arity. For example, the composition of a pair of relations S_1 and S_2, usually denoted by $S_1 ; S_2$, is represented by a three-place relation, $R_;^3$.

Definition 1 $\mathfrak{A} = \langle A; -, \vee, o_{i \in I}^{n_i}\rangle$ *is a* Boolean algebra with operators *(a* **BAO**) *when the equations* (a1)–(a5$_i$) *hold* ($\forall i \in I$ *and* $\forall j\, 1 \leq j \leq n_i$). ($\bot$ *abbreviates* $-(a \vee -a)$, *and* $o_i(\vec{a}, [\ ]_j)$ *indicates that the jth argument has been singled out, while $\vec{a}$ fills the other argument places.* $a, b, c, \ldots \in A$.)

(a1) $a \vee b = b \vee a$ (a2) $(a \vee b) \vee c = a \vee (b \vee c)$
(a3) $-(-a \vee -b) \vee -(-a \vee --b) = a$
(a4$_i$) $o_i(\vec{a}, [b \vee c]_j) = o_i(\vec{a}, [b]_j) \vee o_i(\vec{a}, [c]_j)$
(a5$_i$) $o_i(\vec{a}, [\bot]_j) = \bot$

---

[4] We stress that these are intuitive renderings only. For instance, Rww reads as w is in its own future and in its own past.

On the Origins of Gaggle Theory

Any operation satisfying (a4$_i$) is *additive* in each argument, and with (a5$_i$) added, o_i is *normal*. Of course, in concrete cases, the operators of a **BAO** may interact with each other or may have further properties. For a concrete example, we take an operator $\circ$ that is binary and satisfies two inequations (a6) $(a \circ b) \circ c \leq a \circ (b \circ c)$ and (a7) $(a \circ b) \circ c \leq b \circ (a \circ c)$ (which could be written as equations in a **BAO**). We denote this sample **BAO** as $\mathfrak{A}°$.

Definition 2 *A structure is* $\mathfrak{F} = \langle W, R_\circ^3 \rangle$, *and a model is* $\mathfrak{M} = \langle W, R_\circ^3, v \rangle$, *where* $W \neq \emptyset$, $R_\circ^3 \subseteq W^3$, $v(a) \subseteq W$, $v(a) = [\![a]\!]$ *and* (f1)–(m3) *hold*.

(f1) $\forall w_1, w_2, w_3, w_4 \left(\exists w_5 \left(Rw_1 w_2 w_5 \wedge Rw_5 w_3 w_4 \right) \Rightarrow \right.$
$\left. \exists w_5 \left(Rw_2 w_3 w_5 \wedge Rw_1 w_5 w_4 \right) \right)$

(f2) $\forall w_1, w_2, w_3, w_4 \left(\exists w_5 \left(Rw_1 w_2 w_5 \wedge Rw_5 w_3 w_4 \right) \Rightarrow \right.$
$\left. \exists w_5 \left(Rw_1 w_3 w_5 \wedge Rw_2 w_5 w_4 \right) \right)$[5]

(m1) $[\![-a]\!] = W \setminus [\![a]\!]$ (m2) $[\![a \vee b]\!] = [\![a]\!] \cup [\![b]\!]$
(m3) $[\![a \circ b]\!] = \{\, w_3 \colon \exists w_1, w_2 \,(Rw_1 w_2 w_3 \wedge w_1 \in [\![a]\!] \wedge w_2 \in [\![b]\!]) \,\}$

What we have so far will only guarantee the existence of a *homomorphic representation* for $\mathfrak{A}°$. Thus, we accumulate some further notions.

Definition 3 *Let* $\mathfrak{A}° = \langle A; -, \vee, \circ \rangle$ *be the* **BAO** *above.* $U \subseteq A$ *is an ultrafilter if (i)* $a, b \in U$ *iff* $a \wedge b \in U$ *(i.e.,* U *is a filter) and (ii)* $-a \in U$ *iff* $a \notin U$. *The set of ultrafilters is denoted by* $\mathfrak{U}$. *(iii)* $R_\circ^3 u_1 u_2 u_3$ *iff* $\forall a_1, a_2 ((a_1 \in u_1 \wedge a_2 \in u_2) \Rightarrow a_1 \circ a_2 \in u_3)$ *(where the u's are from* $\mathfrak{U}$).

We only state the following lemmas and a theorem, which are well known, and their proofs are easy or may be found in various publications.

Lemma 1 *For any* $a, b \in A \setminus \{\bot\}$ *in a* **BAO**, *if* $a \not\leq b$, *then there are* $u_a, u_b \in \mathfrak{U}$ *such that* $a \in u_a$, $b \notin u_a$ *and* $b \in u_b$.

Lemma 2 *Let* $R'_\circ$ *be defined as* $R_\circ$ *above, with* $u_1, u_2 \in \mathfrak{F}$ *(where* $\mathfrak{F}$ *is the set of proper filters). If* $R'_\circ u_1 u_2 u_3$ *holds, then there are* u'_1, u'_2 *such that* $u_1 \subseteq u'_1$, $u_2 \subseteq u'_2$ *and* $R_\circ u'_1 u'_2 u_3$ *(where the u''s are from* $\mathfrak{U}$).

Lemma 3 *If* $\mathfrak{A}°$ *is a* **BAO** *as above, then* $R_\circ$ *satisfies* (f1) *and* (f2).

Theorem 2 *Let* $\mathfrak{A}° = \langle A; -, \vee, \circ \rangle$ *be a* **BAO** *as above, and let* $h(a) = \{\, U \in \mathfrak{U} \colon a \in U \,\}$. $h[A]$ *is a concrete* **BAO** *with the operations defined as in* (m1)–(m3) *that is isomorphic to* $\mathfrak{A}°$.

---
[5]We could have written (f1) and (f2) using usual notation for composition of R^3. (f1) would turn into $R(w_1 w_2) w_3 w_4 \Rightarrow Rw_1 (w_2 w_3) w_4$ with the universal closure tacitly assumed.

This isomorphic representation theorem may be viewed as a *completeness theorem* for a logic that has a binary fusion connective of a certain kind on the basis of **TV**. The homomorphic representation theorem similarly parallels the *soundness theorem* for a logic.

The limitation to operators may appear a drastic restriction even if the operators may have further properties. But it is not, because **BA**'s are overly abundant in definable operations. For example, $\Diamond$ which is a unary operator in the Lindenbaum algebra of a normal modal logic, allows one to define $\Box$, but also $\neg\Diamond$ (impossible) and $\Diamond\neg$ (possibly not). Obviously, every operation that can be expressed by a contextual definition in a **BAO** falls under the scope of Theorem 2. However, we have seen in §2.1 that there are operators (or $\Diamond$-like connectives) that cannot be defined from each other on the basis of a **BA**, yet they can be modeled from one relation by switching arguments. **BAO**'s are an archetypical example, where a wide range of operations can be captured by a sole representation theorem. We note that all the operations have *distribution types* and *respect the bounds* too—in the terminology of gaggle theory. (Definitions of these and related notions for different kinds of gaggles may be found in Bimbó and Dunn (2008). See e.g., Definitions 1.3.2, 1.3.18, 2.4.1 and 4.3.13.) At the same time, **BAO**'s are an example where the potential interactions of operations (e.g., through residuation) are not fully exploited in the representation, and in this sense, **BAO**'s are *not* completely general.

2.3 Semantics for intuitionistic logic

Intuitionistic logic (**J**) differs from normal modal logics and **BAO**'s, because its Lindenbaum algebra does not have a **BA** reduct, or in other words, **J** is not simply an extension of **TV**. We only mention one of the several interpretations that have been introduced for **J**, namely, the semantics that originated in Kripke (1965). We assume that the reader is familiar with some formalization of propositional **J**—as an axiomatic system, a sequent calculus, a tableau system or such.

A *frame* (a structure) for **J** is $\mathfrak{F} = \langle U, \sqsubseteq \rangle$, where $U \neq \emptyset$, $\sqsubseteq \subseteq U^2$ and $\sqsubseteq$ is a weak partial order. A *model* is $\mathfrak{M} = \langle U, \sqsubseteq, v \rangle$, where $v(p) = X$ and $X \in \mathcal{P}(U)^{\uparrow}$. ($X \in \mathcal{P}(U)^{\uparrow}$ iff $u' \in X$, whenever $u \sqsubseteq u'$ and $u \in X$.) That is, a propositional variable is mapped by v into a *cone of situations*. Formulas are interpreted by extending v according to clauses (j1)–(j5).

(j1) $[\![\mathcal{A} \wedge \mathcal{B}]\!] = [\![\mathcal{A}]\!] \cap [\![\mathcal{B}]\!]$ $\quad\quad$ (j2) $[\![\mathcal{A} \vee \mathcal{B}]\!] = [\![\mathcal{A}]\!] \cup [\![\mathcal{B}]\!]$
(j3) $[\![\mathcal{A} \to \mathcal{B}]\!] = \{\, u \colon \forall u'(u \sqsubseteq u' \Rightarrow (u' \notin [\![\mathcal{A}]\!] \vee u' \in [\![\mathcal{B}]\!])) \,\}$

(j4) $[\![\neg \mathcal{A}]\!] = \{\, u\colon \forall u'(u \sqsubseteq u' \Rightarrow u' \notin [\![\mathcal{A}]\!])\,\}$ (j5) $[\![\bot]\!] = \emptyset$

$\mathcal{A}$ is *true at u* in $\mathfrak{M}$ if $u \in [\![\mathcal{A}]\!]$; $\mathcal{A}$ is *true in* $\mathfrak{M}$, when $U \subseteq [\![\mathcal{A}]\!]$. Lastly, $\mathcal{A}$ is *valid*, if it is true in all models on all frames for **J**.

The relationship between $\neg \mathcal{A}$ and $\mathcal{A} \to \bot$ is quite clear semantically (and it matches the syntactic definition of $\neg \mathcal{A}$). The crucial clause is (j3), which views $\to_\mathbf{J}$ *almost* as $\supset$, but only in the set of situations that are accessible from the current situation. To facilitate comparison with (j3), we may rewrite 2, the condition for $\supset$, as $[\![\mathcal{A} \supset \mathcal{B}]\!] = \{\, w\colon w \notin [\![\mathcal{A}]\!] \vee w \in [\![\mathcal{B}]\!]\,\}$. Of course, it is well known that $\to_\mathbf{J}$ is very close to $\supset$. The implicational theorems of **TV** that go beyond the implicational theorems of **J** (after the $\to_\mathbf{J}$'s are rewritten into $\supset$'s) are not principal simple type schemas of proper combinators.[6] Or we may note that the sequent calculus LJ results from LK by an uncomplicated structural restriction.

Nevertheless, we may observe that the pattern in the possible world semantics for normal modal logics, which is explicit and general in the representation of **BAO**'s is infringed upon by (j3). We have a binary sentential connective $\to_\mathbf{J}$, but we do not have a ternary relation. Dunn (1995) provided a semantics for **J** along the lines of gaggle theory without taking into consideration potential simplifications. Before recalling how to move back and forth between the two types of semantics, we state the soundness and completeness theorem for the semantics outlined.

Theorem 3 *A formula $\mathcal{A}$ is a theorem of* **J** *iff $\mathcal{A}$ is* valid *on all models.*

We will not give a proof of this theorem here; rather, we sketch the components of the canonical model that would be used for showing the "if" direction of the claim; they also figure into Dunn's completeness theorem.

The Lindenbaum algebra of **J** is a residuated distributive lattice with bottom, and it does not need to be a **BA**. Accordingly, ultrafilters (or equivalently maximally consistent sets of sentences) cannot be used in the representation of such a lattice (or in a model of **J**). The *canonical frame* is $\mathfrak{F}_c = \langle \mathcal{P}, \subseteq \rangle$, where $\mathcal{P}$ is the set of (proper) prime filters. A prime filter P satisfies (i) from Definition 3, (iv) $a, b \in P$ iff $a \vee b \in P$ and (v) $P \neq A$. $\subseteq$ is set inclusion (the prototypical partial order), which may hold between distinct prime filters. The *canonical valuation* is defined as $v(p) = \{\, P \in \mathcal{P}\colon [p] \in P\,\}$. We can decipher all the P's by saying that $v(p)$ is the set of prime filters, in which the equivalence class of p is an element. The following lemmas are helpful in the proof of the completeness theorem.

---

[6]This observation of H. B. Curry is well known; see, e.g., Hindley (1997), Hindley and Seldin (2008) and Bimbó (2012).

Lemma 4 *For any $[\mathcal{A}], [\mathcal{B}]$ in the Lindenbaum algebra of* **J**, *if $[\mathcal{B}] \neq [\bot]$ and $\mathcal{A} \to \mathcal{B}$ is not a theorem of* **J**, *then there are $P_a, P_b \in \mathcal{P}$ such that $[\mathcal{A}] \in P_a$, $[\mathcal{B}] \notin P_a$ and $[\mathcal{B}] \in P_b$. Therefore, $P_b \not\subseteq P_a$.*

Lemma 5 *For any formula $\mathcal{A}$ and prime filter P, $P \in [\![\mathcal{A}]\!]$ iff $[\mathcal{A}] \in P$.*

We briefly recall from Dunn (1995) how a ternary relational semantics for **J** is obtained. A Heyting algebra (exemplified by the algebra of **J**) is residuated, where $\to$ is a residual (indeed, *the* residual) of $\wedge$. The truth condition for $\to$—using a ternary relation—is (j6), and that for $\wedge$ is (j7).

(j6) $[\![\mathcal{A} \to \mathcal{B}]\!] = \{\, u \colon \forall u', u''((Ruu'u'' \wedge u' \in [\![\mathcal{A}]\!]) \Rightarrow u'' \in [\![\mathcal{B}]\!])\,\}$
(j7) $[\![\mathcal{A} \wedge \mathcal{B}]\!] = \{\, u'' \colon \exists u, u'(Ruu'u'' \wedge u \in [\![\mathcal{A}]\!] \wedge u' \in [\![\mathcal{B}]\!])\,\}$

The latter clause suggests that u'' should be a superset of both u and u' on the analogy of $[a), [b) \subseteq [a \wedge b)$. Thus, the ternary relation is defined as $Ruu'u''$ iff $u \sqsubseteq u''$ and $u' \sqsubseteq u''$. Looking at $\wedge$ once more, the identity element is $[\top]$ (i.e., $[\neg\bot]$), which is an element of every prime filter. Thus, a ternary relational frame for **J** is $\mathfrak{F} = \langle U, \sqsubseteq, I, R \rangle$, where $U \neq \emptyset$, $I = U$ and $Ruu'u''$ is defined from the pre-order $\sqsubseteq$ as above. It is easy to see that the truth conditions (j1) and (j7) are equivalent. We quickly run through the proof that (j3) and (j6) are equivalent too. If $u \in [\![\mathcal{A} \to \mathcal{B}]\!]$, and also $Ruu'u''$ and $u' \in [\![\mathcal{A}]\!]$, then by $u' \sqsubseteq u''$, $u'' \in [\![\mathcal{A}]\!]$ follows, because propositions are cones of situations. But $u \sqsubseteq u''$ and $u'' \in [\![\mathcal{A}]\!]$ imply, by (j3), that $u'' \in [\![\mathcal{B}]\!]$, that is, (j6) holds. Now, if we assume $u \in [\![\mathcal{A} \to \mathcal{B}]\!]$, and $u \sqsubseteq u'$ and $u' \in [\![\mathcal{A}]\!]$, then using $u' \sqsubseteq u'$, we have that $Ruu'u'$, and by (j6), $u' \in [\![\mathcal{B}]\!]$.

The ternary modeling may seem only to complicate things. However, that $\to$ is the residual of conjunction $\wedge$ explains the properties of $\to$, and in turn, the properties of R. The residuation between $\wedge$ and $\to$ also implies that all the theorems are equivalent, hence, any and all of them are typified by $\top$.

2.4 Semantics for R-mingle

The logic **RM** is obtained from **R** by adding the mingle axiom (cf. Anderson, Belnap, and Dunn (1992, §R)). **RM** was introduced by Dunn adapting a suggestion of S. McCall (see Dunn (2021)). This logic is often called *semi-relevant* because it has theorems of the form $\mathcal{A} \to \mathcal{B}$, where $\mathcal{A}$ is the negation of a theorem and $\mathcal{B}$ is a theorem. Two instances of a theorem with disjoint sets of propositional variables easily let us create a theorem with implication as its main connective, but no variable partaking in both the antecedent and the consequent. If the variable sharing property is taken to be the hallmark of a relevance logic, then **RM** falls short, because it only satisfies the *weak relevance principle* (cf. Anderson and Belnap (1975, §29.4)). However, **R**-mingle has many pleasant features; in particular, it has a linearly ordered infinite characteristic matrix. The only (non-trivial) linearly ordered **BA** is **2**, the two-element Boolean algebra, and residuated distributive lattices with a least element do not need to be linearly ordered. The semantics that Dunn designed for **RM** toward the end of the 1960s (cf. Dunn (1976b)) seems to follow closely Kripke's terminology and notation; however, those similarities turn out to be quite superficial. Some of the novel properties of Dunn's semantics include: (1) The semantic uses a *generated model* (in the contemporary sense of the term in the modal logic literature). (2) The semantics is 3-*valued*, moreover, the three truth values are $\{T\}$, $\{F\}$ and $\{T, F\}$, that is, the non-empty subsets of the "usual" set of truth values. (3) The semantic uses a *distinguished situation*—like Kripke's semantics, but the distinguished situation cannot be an arbitrary situation—unlike in Kripke's semantics. (4) The frame is *linearly ordered*, which is not stipulated in the semantics for normal modal logics in general or in the semantics for **J**.

Definition 4 *A frame for* **RM** *is* $\mathfrak{F} = \langle U, \iota, \leq \rangle$, *where* $\iota \in U$, $\leq \,\subseteq U^2$ *and* $\leq$ *is reflexive, transitive and connected. For ease of use,* $\leq$ *is stipulated to be anti-symmetric with ι being the least element in U. A model is* $\mathfrak{M} = \langle U, \iota, \leq, v \rangle$, *where* $v\colon \mathbb{P} \times U \longrightarrow \{\{T\}, \{F\}, \{T, F\}\}$ *satisfying* hereditariness, *that is,* (h) *if* $u \leq u'$, *then* $v(p, u) \subseteq v(p, u')$. v *is extended to compound formulas according to* (1)–(4) *(below).*

The condition (h) is stipulated for $p \in \mathbb{P}$ (i.e., propositional variables). To contrast this with the condition in a model for **J**, we express the former from page 24 in a way similar to (h). Thus, (h$_\mathbf{J}$) says that if $u \sqsubseteq u'$ then $u \in v(p)$ implies $u' \in v(p)$, or in other words, if $u \sqsubseteq u'$ then $v'(p, u) = \{T\}$ implies $v'(p, u') = \{T\}$ (where using v', we transformed v into a binary function in an obvious way). In the semantics of **J**, $\{T, F\}$ cannot be the value for any p in any situation; hence, $v'(p, u') = \{T, F\}$ in the consequent is not possible. Essentially for the same reason, $v'(p, u') = \{F\}$ is not stipulated when $u \sqsubseteq u'$ and $v'(p, u) = \{F\}$. To put it concisely, for **J**, the perpetuation of truth is required along the accessibility relation, whereas for **RM** both truth and falsity are upheld moving forward along the linear order.

The view of the three truth values as sets allows for a straightforward extension of v as follows.

(1) $T \in v(\mathord{\sim}\mathcal{A}, u)$ iff $F \in v(\mathcal{A}, u)$;

$F \in v(\sim\mathcal{A}, u)$ iff $T \in v(\mathcal{A}, u)$;
(2) $T \in v(\mathcal{A} \wedge \mathcal{B}, u)$ iff $T \in v(\mathcal{A}, u)$ and $T \in v(\mathcal{B}, u)$;
  $F \in v(\mathcal{A} \wedge \mathcal{B}, u)$ iff $F \in v(\mathcal{A}, u)$ or $F \in v(\mathcal{B}, u)$;
(3) $T \in v(\mathcal{A} \vee \mathcal{B}, u)$ iff $T \in v(\mathcal{A}, u)$ or $T \in v(\mathcal{B}, u)$;
  $F \in v(\mathcal{A} \vee \mathcal{B}, u)$ iff $F \in v(\mathcal{A}, u)$ and $F \in v(\mathcal{B}, u)$;
(4) $T \in v(\mathcal{A} \to \mathcal{B}, u)$ iff $\forall u'(u \leq u' \Rightarrow$
  $(T \in v(\mathcal{A}, u') \Rightarrow T \in v(\mathcal{B}, u'). \wedge . F \in v(\mathcal{B}, u') \Rightarrow F \in v(\mathcal{A}, u')))$;
  $F \in v(\mathcal{A} \to \mathcal{B}, u)$ iff $T \notin v(\mathcal{A} \to \mathcal{B}, u)$ or
  $T \in v(\mathcal{A}, u)$ and $F \in v(\mathcal{B}, u)$.

First, we note that the T and F clauses are *independent* from each other, and even in the case of the extensional connectives ($\sim$, $\wedge$ and $\vee$), neither line may be omitted. Second, the T clause for $\to$ starts similarly to the $\to$ clause in **J**, but here the falsity of the consequent must imply the falsity of the antecedent too. It may be also notable that only the T condition for $\to$ can shift the evaluation to a new situation.

The *truth of* $\mathcal{A}$ at a situation u means that the value of the formula is $\{T\}$ or $\{T, F\}$, that is, $T \in v(\mathcal{A}, u)$. *Truth in a model* means truth at ι, and *validity* obtains when a formula is true in all models. Dunn (1976b) proved the following soundness and completeness theorems.

Theorem 4 *If $\mathcal{A}$ is a theorem of* **RM**, *then $\mathcal{A}$ is* valid, *and* vice versa.

We outline the canonical model and point out some of its properties. The *canonical frame* is defined with respect to a prime theory T_0 (or in the Lindenbaum algebra, a prime filter) that contains every theorem of **RM**. U_c is the set of all prime theories that extend T_0 (and contain all **RM** theorems). T_0 is ι_c and $\leq_c$ is $\subseteq$. Obviously, $\langle U_c, \iota_c, \leq_c \rangle$ is a frame for **RM**. (Connectedness follows from the chain theorem $(\mathcal{A} \to \mathcal{B}) \vee (\mathcal{B} \to \mathcal{A})$.) The canonical valuation is defined by a conjunctive condition, namely, $T \in v_c(p, u)$ iff $p \in u$, and $F \in v_c(p, u)$ iff $\sim p \in u$.

The selection of a prime theory ι_c that contains all the theorems of **RM** is motivated by the definition of the frame. But it also shows that in **RM**—as in relevance logics typically—not all theorems are equal, which means that the top element of the Lindenbaum algebra of the logic (if there is one) does not stand for (or implies) all theorems.

Now, we turn to sketching another semantics for **RM**, which is a step closer to what would result from an application of gaggle theory, and more in line with the Meyer–Routley semantics (though we diverge from the usual presentation of the latter). A *frame* is $\mathfrak{F} = \langle U, \iota, R, * \rangle$, where $\iota \in U$, $*: U \longrightarrow U$, $u^{**} = u$, $Ru'us \Rightarrow Ru's^*u^*$ and R satisfies (f3)–(f8).

(f3) $R(uu')ss' \Leftrightarrow Ru'(us)s'$;   (f4) $R(uu')ss' \Rightarrow R(us)u's'$;
(f5) $R\iota uu$;   (f6) $(R\iota u'u \wedge R\iota s's \wedge R\iota u''s'' \wedge Rusu'') \Rightarrow Ru's's''$;
(f7) $Russ' \Rightarrow R(us)ss'$;   (f8) $Ruu's \Rightarrow (R\iota us \vee R\iota u's)$.

A pre-order relation can be recovered as $u \sqsubseteq u'$ iff $R\iota uu'$. Then the last condition (which "matches" the mingle axiom) may be written as $Ruu's \Rightarrow$ $(u \sqsubseteq s \vee u' \sqsubseteq s)$. We note that taking a single logical situation, namely, ι is justified by the fact that the Lindenbaum algebra of **RM** is linearly ordered. A model is obtained by adding v, which maps p into a cone of situations (with respect to $\sqsubseteq$). It should be immediately clear that this semantics is 2-valued, because we have not mentioned any truth values. v is extended to compound formulas by intersection and union for $\wedge$ and $\vee$, respectively. $\mathcal{A} \to \mathcal{B}$ is evaluated as in (j6) (with $\to$ taken to be the implication of **RM**). The remaining clauses are (m4) and (m5).

(m4) $[\![t]\!] = [\iota)$ (m5) $[\![\sim \mathcal{A}]\!] = \{u : u^* \notin [\![\mathcal{A}]\!]\}$

It seems fair to say that Dunn's 3-valued semantics is more *elegant*. The frame is a kind of structure that is quite familiar to us; e.g., it is exemplified by $\mathbb{N}$ (which of course, brings additional properties with itself). The truth and falsity conditions for $\sim$, $\wedge$ and $\vee$ surely look familiar. The conditions for $\to$ are ingenious, but straightforward—except perhaps, when the implication is assigned F, because it's not true. (Dunn expressed a certain dissatisfaction with this disjunct and called it the "escape clause.") On the other hand, the ternary relational semantics treats **RM** as yet another intensional logic. The multiple conditions on R stem from the fact that **RM** extends **R**; only the last condition is specific to mingle (over **R**). Indeed, Meyer, as soon as he specified the conditions for R, commented with vexation on their number and somewhat complicated character. However, this complexity is the price for approximating $\supset$ as much as relevantly possible (and perhaps, it is a price for generality too). The *flexibility* of the ternary relational semantics for various relevance logics such as **B**, **T**, **E** and **R** was undoubtedly an impetus for Dunn's formulation of gaggle theory.

2.5 Semantics for relevance logics

The 3-valued semantics for **RM** does not seem to be easily adaptable to some other relevance logics, especially, to the main relevance logics we have just mentioned (i.e., **B**, **T**, **E** and **R**) that do not contain the mingle axiom.[7] Another way to think about a concrete semantics is by having operations on a set of objects.[8] However, the latter idea does not work for the semantics of relevance logics just as it did not work for the semantics of normal modal logics; the natural operation on prime filters does not yield a prime filter.

A semantics that uses a ternary relation for the modeling of $\to$ and $\circ$ was worked out in detail and published by Routley and Meyer (1972a, 1972b, 1973). A leftover from the operational approach is the modeling of $\sim$ from an operation (cf. Dunn (1966, 1986)). A different combination of operations and relations is used in the semantics in Fine (1974), which in effect, turns out to be equivalent to the Meyer–Routley semantics.[9]

The formulation of the ternary relational semantics seems to have propelled the creation of gaggle theory. Dunn and Meyer both worked at Indiana University (in Bloomington, IN), when Meyer—inspired by an idea in Routley's big manuscript—worked out the relational semantics for $\mathbf{R}^\circ$ in a form that is very close to its later presentations (cf. Bimbó, Dunn, and Ferenz (2018)). Logics in which there is a conjunction and disjunction that distribute over each other are well behaved (from the point of view of their semantics), and the main relevance logics (in their full vocabulary) are among those (just as **J**). The first notion of a *gaggle* introduced in Dunn (1991) incorporates a *distributive lattice* as the living quarters for a family of operations, and has been called a *distributive gaggle* afterward (cf. Bimbó and Dunn (2008)). To illustrate both the Meyer–Routley semantics and a concrete (multi-)gaggle, we will use $\mathbf{T}^{\circ t}$ and its algebra. An axiomatization of ticket entailment may be found in Anderson et al. (1992, §R); we assume familiarity with this logic.

Definition 5 *A $\mathfrak{G}_\mathbf{T}$ gaggle is an algebra $\langle A; \wedge, \vee, \sim, t, \circ, \to \rangle$ of similarity type $\langle 2, 2, 1, 0, 2, 2 \rangle$, where* (g1)–(g6) *hold.*

(g1) $\langle A; \wedge, \vee, \sim \rangle$ *is a De Morgan lattice;*
(g2) $\langle A; \wedge, \vee, t, \circ, \to \rangle$ *is a lattice ordered groupoid (with $\circ$) with left identity ($t \circ a = a$) and with right residual ($a \circ b \leq c$ iff $a \leq b \to c$);*
(g3) $(a \circ b) \circ c \leq a \circ (b \circ c)$; (g4) $(a \circ b) \circ c \leq b \circ (a \circ c)$;
(g5) $a \circ b \leq (a \circ b) \circ b$; (g6) $a \circ b \leq c$ iff $a \circ \sim c \leq \sim b$.

---

[7]Some adaptations work well though. See Dunn (1976a) and Bimbó and Dunn (in press).
[8]Bimbó and Dunn (2017) provides an overview of some of the early work toward a set-theoretical semantics for relevance logics by several logicians, e.g., Urquhart (1972). We will not repeat that history here; rather, we focus on the Meyer–Routley semantics.
[9]Semantics that are duals of the Meyer–Routley semantics were defined for **T** and **E** in Bimbó (2007) and Bimbó (2009); the latter also includes a topological characterization.

On the Origins of Gaggle Theory

We called $\mathfrak{G}_{\mathbf{T}}$ a gaggle, but it is really two gaggles and a constant integrated into one algebra. The constant t is connected to the $\circ$ gaggle (which includes $\to$) and this interacts with the $\sim$ gaggle (which is a component of the De Morgan lattice). For any $\mathcal{A}$ that is a theorem of $\mathbf{T}^{\circ t}$, the formula $t \to \mathcal{A}$ is provable. Thus, we may think of $\mathfrak{G}_{\mathbf{T}}$ as a matrix, with $D = \{\, a \colon t \leq a\,\}$. In the modeling of $\circ$ and $\to$, we follow the Meyer–Routley semantics, but for t and $\sim$ we make some modifications.

Definition 6 *A frame for* $\mathfrak{G}_{\mathbf{T}}$ *is* $\mathfrak{F} = \langle U, \sqsubseteq, I, R_\circ, R_\sim \rangle$, *where* $I \neq \emptyset$, $I \subseteq U$, $\sqsubseteq$ *is a preorder on* U, $R_\circ \subseteq U^3$, $R_\sim \subseteq U^2$ *and* (f1)–(f7) *also hold.*

(f1) $(R_\circ uu'u'' \wedge s \sqsubseteq u \wedge s' \sqsubseteq u' \wedge u'' \sqsubseteq s'') \Rightarrow R_\circ ss's''$ (*i.e.*, $R_\circ\!\downarrow\!\downarrow\!\uparrow$);

(f2) $u \sqsubseteq u' \Leftrightarrow \exists \iota \in I \, R_\circ \iota u u'$; $I \in \mathcal{P}(U)^\uparrow$; $R_\sim\!\uparrow\!\uparrow$;

(f3) $\exists u \, (\neg R_\sim uu' \wedge \forall u'' (\neg R_\sim u''u \Rightarrow u'' \sqsubseteq u'))$;

(f4) $(R_\circ uu'u'' \wedge \neg R_\sim su'') \Rightarrow \exists s', s'' \, (R_\circ uss' \wedge \neg R_\sim s''s' \wedge u' \sqsubseteq s'')$;

(f5) $\neg R_\sim u'u'' \Rightarrow \exists s, s', s'' \, (R_\circ u''ss' \wedge u' \sqsubseteq s \wedge u' \sqsubseteq s'' \wedge \neg R_\sim s''s')$;

(f6) $R_\circ(uu')ss' \Rightarrow R_\circ u(u's)s'$; $R_\circ(uu')ss' \Rightarrow R_\circ u'(us)s'$;

(f7) $R_\circ uu's \Rightarrow R_\circ(uu')u's$.

The frame is defined to have a pre-order, which makes this frame somewhat similar to that for **J**. But now $\sqsubseteq$ is definable from $R_\circ$ and I, rather than R being definable from $\sqsubseteq$ as in the case of **J**. This is explained by the fact that in $\mathbf{T}^{\circ t}$ implication is not a residual of $\wedge$, and $\sqsubseteq$ is linked to provable implications. The ternary relation in the semantics of **J** seemed almost like a vapid complication, though we made some pertinent observations using R. Here the use of a binary relation $R_\sim$ instead of a unary operation is a similar intricacy; the operation could be denoted by $*$, as at the end of §2.4. If we let u^* to be s, then R^*us is definable as $\neg R_\sim us \wedge \neg \exists s' \, (s \neq s' \wedge \neg R_\sim us' \wedge s \sqsubseteq s')$. It so happens that in the Lindenbaum algebra of $\mathbf{T}^{\circ t}$ such an s always exists, moreover, it is unique; these properties support the use of an operation. (The inequations in (g3) and (g4) are the previous (a6) and (a7) in §2.4, and the two conditions in (f6) are the same as (f1) and (f2) in Definition 2.)

Definition 7 *A model for* $\mathfrak{G}_{\mathbf{T}}$ *is* $\mathfrak{M} = \langle U, \sqsubseteq, I, R_\circ, R_\sim, v \rangle$, *where the frame is as above and* $v \colon \mathbb{P} \longrightarrow \mathcal{P}(U)^\uparrow$, *which is extended to all formulas according to* (m1)–(m7).

(m1) $[\![p]\!] = v(p)$ (m2) $[\![t]\!] = \{\, u \colon \exists \iota \in I \, \iota \sqsubseteq u\,\}$

(m3) $[\![\mathcal{A} \wedge \mathcal{B}]\!] = [\![\mathcal{A}]\!] \cap [\![\mathcal{B}]\!]$ (m4) $[\![\mathcal{A} \vee \mathcal{B}]\!] = [\![\mathcal{A}]\!] \cup [\![\mathcal{B}]\!]$

(m5) $[\![\sim\!\mathcal{A}]\!] = \{\, u \colon \forall u'(u' \in [\![\mathcal{A}]\!] \Rightarrow R_\sim u'u)\,\}$

(m6) $[\![\mathcal{A} \circ \mathcal{B}]\!] = \{\, u'' \colon \exists u, u' (R_\circ uu'u'' \wedge u \in [\![\mathcal{A}]\!] \wedge u' \in [\![\mathcal{B}]\!])\,\}$
(m7) $[\![\mathcal{B} \to \mathcal{C}]\!] = \{\, u \colon \forall u', u'' ((R_\circ uu'u'' \wedge u' \in [\![\mathcal{B}]\!]) \Rightarrow u'' \in [\![\mathcal{C}]\!])\,\}$

A formula $\mathcal{A}$ is *true at the situation* u (in some $\mathfrak{M}$), when $u \in [\![\mathcal{A}]\!]$. The *truth* of $\mathcal{A}$ in a model means that $\forall \iota \in I\, \iota \in [\![\mathcal{A}]\!]$, that is, $[\![t]\!] \subseteq [\![\mathcal{A}]\!]$. In Dunn's three-valued semantics for **RM**, stipulating that ι was the least element of U had the effect of limiting U to *logical situations*, which is similar to requiring $U = I$ in the case of **J**. However, here we did not assume that all situations are logical, neither have we stated that I is the principal cone generated by a particular logical situation. (In some presentations of the Meyer–Routley semantics occasionally one distinguished situation is selected, which is denoted by 0; we do not follow that track here.) *Validity* means truth in every model of a frame for $\mathfrak{G}_\mathbf{T}$. The proof of the following is easy and we do not include the details here.

Lemma 6 (Hereditary property) *For all formulas $\mathcal{A}$, $[\![\mathcal{A}]\!] \in \mathcal{P}(U)^\uparrow$.*

This lemma means that truth is retained along the $\sqsubseteq$ relation (which is not the accessibility relation as in **J**, only a special part of it). The import of the lemma is that propositions (i.e., interpretations of formulas) are located among the upward closed sets of situations. To this extent the lemma is similar to the hereditariness lemma in the semantics of **J**.

What we have so far suffices for soundness. For completeness, we outline the definition of the canonical frame and that of the canonical model.

Definition 8 *The* canonical frame *for $\mathfrak{G}_\mathbf{T}$ is $\mathfrak{F}_c = \langle U_c, \subseteq, I_c, R_\circ, R_\sim \rangle$, where $U_c = \mathcal{P}$, $I_c = \{\, P \in \mathcal{P} \colon [t] \subseteq P\,\}$ and $R_\sim, R_\circ$ are as in* (c1)–(c2).

(c1) $R_\sim uu' \Leftrightarrow \exists a\, (a \in u \wedge {\sim} a \in u')$
(c2) $R_\circ uu'u'' \Leftrightarrow \forall a, b\, ((a \in u \wedge b \in u') \Rightarrow a \circ b \in u'')$

The canonical model *for $\mathfrak{G}_\mathbf{T}$ is $\mathfrak{M}_c = \langle \mathfrak{F}_c, v_c \rangle$, where $\mathfrak{F}_c$ is the canonical frame and $v_c([\mathcal{A}]) = \{\, P \in \mathcal{P} \colon [\mathcal{A}] \in P\,\}$.*

To get to Theorem 5, it is convenient to prove certain claims as lemmas, which we only list here. First, the components of $\mathfrak{F}_c$ are of the declared types, however, it is far from obvious that they have all the required properties, especially, that $R_\circ$ and $R_\sim$ satisfy (f3)–(f7). In establishing the properties of $R_\sim$ and $R_\circ$, it is useful to prove versions of the *squeeze lemma*. The latter then may be utilized in the proof that v_c is a homomorphism. The canonical situations are prime filters, hence, it is sufficient to appeal to a well-known result from lattice theory about separation to see that v_c is injective, that is, an isomorphism. We simply state adequacy. The proofs may be found or can be pieced together from results in some of the publications cited.

Theorem 5 *For any formula* $\mathcal{A}$, $\vdash_{\mathbf{T}^{\circ\iota}} \mathcal{A}$ *iff on any frame* $\mathfrak{F}$ *for* $\mathfrak{G}_{\mathbf{T}}$, $\vDash_{\mathfrak{F}} \mathcal{A}$.

3 Concluding remarks

I attempted to reconstruct the conceptual components that likely influenced the formulation of gaggle theory. The sources for the reconstruction were Dunn's publications related to gaggle theory and a more comprehensive view of Dunn's research including his other publications and research talks.

Dunn (1966) algebraized $\mathbf{R}^t$ (and $\mathbf{E}$ too). Although much of the research in relevance logics was guided by Anderson (1963) at the time, with a focus on proving (propositional) $\mathbf{R}$ and $\mathbf{E}$ decidable, Dunn formulated the first relational semantics for an intensional logic (other than modal logics and $\mathbf{J}$) in the late 1960s (published as Dunn (1976b, 1976c)). He continued to publish on algebraic semantics and results for intensional logics (propositional and quantified) as well as on other aspects of intensional logics. However, after the invention of the Meyer–Routley semantics for relevance logics, Dunn published Dunn (1976d) and half a decade later Dunn (1982), which are alternative relational semantics for some logics.

Dunn gave over 200 research talks in his career; it seems that the first *gaggle talk* was delivered in Canada, in 1979, at the University of Victoria that was entitled "Generalized Representation and Completeness Results." In 1983, Dunn toured Australia, and gave several talks on relevance logic and other topics. The following year he gave a talk at the Carnegie–Mellon University, which mentions Galois in its title "A Uniform Treatment of Implication and Negation through Residuation and Galois Connections." Already in 1983, in a research proposal, Dunn stated that most of the representation results for a range of logics had been obtained.

In sum, the emergence of *Generalized Galois Logics* (or *gaggle theory*) can be safely dated to about a decade or so earlier than the publication of Dunn (1991), which started a series of papers on gaggle theory. The delay can be attributed to the abundance of publications by Dunn during this period—such as a series of papers on relevant predication, a co-authored book, a co-edited book, a chapter on relevance logic in a handbook (which became a standard reading in the area), and work on another co-authored book. Dunn also co-authored a short paper Dunn and Hellman (1986) on probability theory during the decade. Somewhat surprisingly, this paper—which is not a paper in logic per se, rather an application of logic—turned out to be Dunn's most widely known and read publication at the time. He received requests for offprints of this paper from all over the world and from people way outside of academia.

Gaggle theory can be seen as an *overarching approach* to propositional intensional logics that starts with an axiomatic calculus or some other proof system, then moves through algebraization to a set-theoretic semantics. Generalized Galois logics, including its development in Dunn and Hardegree (2001), Bimbó and Dunn (2008) and many other publications, proved to be exceptionally fruitful. However, it is worth mentioning that gaggle theory is a relatively modest part of Dunn's overall logical research.

References

Anderson, A. R. (1963). Some open problems concerning the system E of entailment. *Acta Philosophica Fennica*, *16*, 9–18.

Anderson, A. R., & Belnap, N. D. (1975). *Entailment: The Logic of Relevance and Necessity* (Vol. I). Princeton: Princeton University Press.

Anderson, A. R., Belnap, N. D., & Dunn, J. M. (1992). *Entailment: The Logic of Relevance and Necessity* (Vol. II). Princeton: Princeton University Press.

Bimbó, K. (2007). Relevance logics. In D. Jacquette (Ed.), *Philosophy of Logic* (Vol. 5 of *Handbook of the Philosophy of Science*, pp. 723–789). Amsterdam: Elsevier (North-Holland).

Bimbó, K. (2009). Dual gaggle semantics for entailment. *Notre Dame Journal of Formal Logic*, *50*(1), 23–41.

Bimbó, K. (2012). *Combinatory Logic: Pure, Applied and Typed*. Boca Raton: CRC Press.

Bimbó, K. (2020). The development of decidability proofs based on sequent calculi. In A. Rezuş (Ed.), *Contemporary Logic and Computing* (Vol. 1 of *Landscapes in Logic*, pp. 5–37). London: College Publications.

Bimbó, K., & Dunn, J. M. (2008). *Generalized Galois Logics: Relational Semantics of Nonclassical Logical Calculi* (Vol. 188 of *CSLI Lecture Notes*). Stanford: CSLI Publications.

Bimbó, K., & Dunn, J. M. (2017). The emergence of set-theoretical semantics for relevance logics around 1970. *IFCoLog Journal of Logics and*

Their Applications, *4*(3), 557–589. (Special Issue: Proceedings of the Third Workshop, 16-17 May 2016, Edmonton, Canada)

Bimbó, K., & Dunn, J. M. (in press). Entailment, mingle and binary accessibility. In Y. Weiss & R. Padro (Eds.), *Saul A. Kripke on Modal Logic*. Cham: Springer Nature.

Bimbó, K., Dunn, J. M., & Ferenz, N. (2018). Two manuscripts, one by Routley, one by Meyer: The origins of the Routley–Meyer semantics for relevance logics. *Australian Journal of Logic*, *15*(2), 171–209.

Dunn, J. M. (1966). *The Algebra of Intensional Logics* (Doctoral dissertation). University of Pittsburgh, Pittsburgh. (Published as Vol. 2 in the *Logic PhDs* series by College Publications, London, UK, 2019.)

Dunn, J. M. (1976a). Intuitive semantics for first-degree entailments and 'coupled trees'. *Philosophical Studies*, *29*(3), 149–168.

Dunn, J. M. (1976b). A Kripke-style semantics for R-mingle using a binary accessibility relation. *Studia Logica*, *35*(2), 163–172.

Dunn, J. M. (1976c). Quantification and RM. *Studia Logica*, *35*(3), 315–322.

Dunn, J. M. (1976d). A variation on the binary semantics for R-Mingle. *Relevance Logic Newsletter*, *1*(2), 56–67.

Dunn, J. M. (1982). A relational representation of quasi-Boolean algebras. *Notre Dame Journal of Formal Logic*, *23*(4), 353–357.

Dunn, J. M. (1986). Relevance logic and entailment. In D. Gabbay & F. Guenthner (Eds.), *Handbook of Philosophical Logic* (1st ed., Vol. 3, pp. 117–224). Dordrecht: D. Reidel.

Dunn, J. M. (1991). Gaggle theory: An abstraction of Galois connections and residuation, with applications to negation, implication, and various logical operators. In J. van Eijck (Ed.), *Logics in AI: European Workshop JELIA '90* (pp. 31–51). Berlin: Springer.

Dunn, J. M. (1995). Gaggle theory applied to intuitionistic, modal and relevance logics. In I. Max & W. Stelzner (Eds.), *Logik und Mathematik. Frege-Kolloquium Jena 1993* (pp. 335–368). Berlin: W. de Gruyter.

Dunn, J. M. (2021). R-Mingle is nice, and so is Arnon Avron. In O. Arieli & A. Zamansky (Eds.), *Arnon Avron on Semantics and Proof Theory of Non-classical Logics* (Vol. 21 of *Outstanding Contributions to Logic*, pp. 141–165). Cham: Springer Nature.

Dunn, J. M., & Hardegree, G. M. (2001). *Algebraic Methods in Philosophical Logic* (Vol. 41 of *Oxford Logic Guides*). Oxford: Oxford University Press.

Dunn, J. M., & Hellman, G. (1986). Dualling: A critique of an argument of Popper and Miller. *The British Journal for the Philosophy of Science*, *37*, 220–223.

Fine, K. (1974). Models for entailment. *Journal of Philosophical Logic*, *3*(4), 347–372.

Hindley, J. R. (1997). *Basic Simple Type Theory* (Vol. 42 of *Cambridge Tracts in Theoretical Computer Science*). Cambridge: Cambridge University Press.

Hindley, J. R., & Seldin, J. P. (2008). *Lambda-calculus and Combinators, an Introduction*. Cambridge: Cambridge University Press.

Jónsson, B., & Tarski, A. (1951–52). Boolean algebras with operators, I and II. *American Journal of Mathematics*, *73 & 74*, 891–939, 127–162.

Kripke, S. A. (1959). A completeness theorem in modal logic. *Journal of Symbolic Logic*, *24*(1), 1–14.

Kripke, S. A. (1963). Semantical analysis of modal logic I. Normal modal propositional calculi. *Zeitschrift für mathematische Logik und Grundlagen der Mathematik*, *9*, 67–96.

Kripke, S. A. (1965). Semantical analysis of intuitionistic logic I. In J. N. Crossley & M. A. E. Dummett (Eds.), *Formal Systems and Recursive Functions. Proceedings of the Eighth Logic Colloquium* (pp. 92–130). Amsterdam: North-Holland.

Look, B. C. (2016). Leibniz's theories of necessity. In M. Cresswell, E. Mares, & A. Rini (Eds.), *Logical Modalities from Aristotle to Carnap. The story of Necessity* (pp. 194–217). Cambridge: Cambridge University Press.

Routley, R., & Meyer, R. K. (1972a). The semantics of entailment – II. *Journal of Philosophical Logic*, *1*, 53–73.

Routley, R., & Meyer, R. K. (1972b). The semantics of entailment – III. *Journal of Philosophical Logic*, *1*, 192–208.

Routley, R., & Meyer, R. K. (1973). The semantics of entailment. In H. Leblanc (Ed.), *Truth, Syntax and Modality.* (pp. 199–243). Amsterdam: North-Holland.

Urquhart, A. (1972). Semantics for relevant logic. *Journal of Symbolic Logic*, *37*(1), 159–169.

Katalin Bimbó
University of Alberta, Department of Philosophy
Canada
E-mail: bimbo@ualberta.ca

Simple Semantics for Logics of Indeterminate Epistemic Closure

COLIN R. CARET

Abstract: According to Jago (2014a), logical omniscience is really part of a deeper paradox. Jago develops an epistemic logic with principles of indeterminate closure to solve this paradox, but his official semantics is difficult to navigate, it is motivated in part by substantive metaphysics, and the logic is not axiomatized. In this paper, I simplify this epistemic logic by adapting the hyperintensional semantic framework of Sedlár (2021). My first goal is metaphysical neutrality. The solution to the epistemic paradox should not require appeal to a metaphysics of truth-makers, situations, or impossible worlds, by contrast with Jago's official semantics. My second goal is to elaborate on the proof theory. I show how to axiomatize a family of logics with principles of indeterminate epistemic closure.

Keywords: knowledge, closure, paradox, omniscience, hyperintensionality

1 Introduction

On the face of it, logical omniscience is just an artifact of certain semantic techniques (see, e.g. Hintikka 1962). In order to avoid this problem, we just need to find a semantics for knowledge that is not so crude.

If knowledge is not closed under all logical consequences, then a natural follow-up question is whether it is closed under some *restricted class* of logical consequences.[1] We might be tempted to think about this as follows. Some logical consequences are obvious to all epistemic subjects, in virtue of their capacity for rational thought. Call them *trivial* consequences. The inference from "φ and ψ" to its conjunct φ could be a plausible example of triviality. Knowing the individual conjunct seems like it is part of knowing the whole conjunction.[2] It just comes along for free. Although it is difficult to precisely define triviality, the *primitive rules* of proof theory seem to fall into this category because they are the simplest deductive inferences.

---

[1] Cf. Duc (1995) and Jago (2006) on 'logical ignorance'.
[2] Yablo (2014) and Hawke, Özgün, and Berto (2020) defend this claim.

According to Jago (2014a), however, this reveals that logical omniscience is actually part of a deeper problem. On the one hand, we want to deny that knowledge is closed under all logical consequences. On the other hand, we want to assert that knowledge is 'trivially closed', i.e. it is closed under consequences that are derivable by a single use of any primitive rule. The problem is that this is paradoxical: any logical consequence is derivable by a chain of primitive rules, so trivial closure implies full closure.

Jago's solution is a theory of Indeterminate Epistemic Closure (IEC), viz. the view that knowledge closure is only partially determinate and trivial consequences are borderline cases where this determinacy breaks down. On this view, the following principle should hold in an epistemic logic enhanced with operator Δ to be read 'it is determinate that…' (Jago, 2014a, p.251).

(IEC) If $\varphi_1, \ldots, \varphi_n \vDash \psi$ is trivial, $\Delta K\varphi_1, \ldots, \Delta K\varphi_n \vDash \neg\Delta\neg K\psi$

We can gloss IEC as follows: with respect to any trivial inference, it is impossible for 'the break down of knowledge closure' to be determinate. More carefully, it is impossible for a subject to determinately know the premises and determinately fail to know the conclusion of a trivial inference. This notion of determinacy is supposed to be something like the notion used in discussions of vagueness.[3] For Jago, the philosophical importance of this concept largely comes out as a norm of assertion: "…we can never rationally assert that such-and-such is an epistemic oversight…Such cases are always indeterminate cases and as such do not rationally support assertions about them in the way that clear cases do." (Jago, 2014a, p.19)

On this view, when knowledge closure *does* break down over a trivial inference step, it is unassertible *that* it breaks down there. This yields a diagnosis of the paradox as follows. Assuming that the role of triviality is captured by IEC, we should recognize that for any trivial inference, it is indeterminate that knowledge closure breaks down at that step. We might then assert: closure does not break down at that step! This is a mistake, but an easy one to make because this can sound very similar to IEC itself.[4] Once we fall prey to this mistake, however, it compels us to think that knowledge is trivially closed and thereby draws us into the paradox.[5]

---

[3]Where it often used to define borderline cases, i.e. where it is *borderline* φ if, and only if, it is neither determinately φ nor determinately $\neg\varphi$.

[4]The mistake consists in sliding from a claim of the form $\neg\Delta\chi$ that says there is a certain *failure of assertibility*, to a claim of the form $\Delta\neg\chi$ which says there is *an assertible failure*.

[5]Another option to consider is a non-classical approach to K, where the degree of truth of $K\varphi$ is low but non-zero for any conclusion φ of a trivial inference. This would model

Simple Semantics for Logics of Indeterminate Epistemic Closure

In this paper, I will explore a family of IEC logics with two main goals. My first goal is metaphysical neutrality. The problem and solution described above made perfectly good sense without any metaphysical assumptions. In particular, it did not seem to involve any mention of entities such as impossible worlds (as used in Jago's semantics). I will show that a simple semantics is available without appeal to substantive metaphysical categories. This is not because I believe that impossible worlds are problematic in principle, but simply because it seems to me that the formal solution to the paradox is conceptually independent of our metaphysical commitments.

My second goal is to elaborate on the proof theory. Although some properties of the knowledge operator and determinacy operator are clear from Jago's exposition, he has not given a systematic proof theory for IEC logics. I will take a step in that direction. To achieve these ends, I will leverage some recent work on hyperintensional semantics by Sedlár (2021). In the next section, I present an overview of this framework.

2 Sphere models for determinate knowledge

What are the target properties of this epistemic logic? I will start with some negative properties. For one thing, knowledge ought not be closed under replacement of necessary equivalents, i.e. it is a hyperintensional context. Similar intuitions to those that lead us to reject omniscience in the first place should also lead us to reject closure under logical equivalence. For another thing, determinate truth ought to be stronger than 'mere truth'.

- $\varphi \leftrightarrow \psi \nvDash K\varphi \leftrightarrow K\psi$

- $\varphi \nvDash \Delta\varphi$[6]

As for positive properties: both operators are factive, and the determinacy operator is a normal modality (e.g. tautologies are determinately true).

- $K\varphi \vDash \varphi$

- $\Delta\varphi \vDash \varphi$

---

indeterminacy from the metalanguage, instead of adding a marker for determinacy to the classical object-language. Thanks to the editor for this thought.

[6]This is one respect in which the present treatment of determinacy differs from that of supervaluationist logics (see, e.g. Fine 1975). There is also a subtle conceptual difference in that we have in mind an objective rather than subjective kind of determinacy.

- $\models \varphi$ implies $\models \Delta\varphi$

In the semantics below, I will follow Jago's lead and treat the determinacy operator as a strong normal modality like the S5 necessity operator, but in principle it could be understood as some kind of weaker operator.

To articulate an elegant semantics that combines these two operators, I will adapt the hyperintensional semantics of Sedlár (2021). Every sentence has both a fine-grained content (FGC) and a coarse-grained content (CGC) in this semantic framework. These divide the traditional work of a *proposition*. The CGC of a sentence is simply a set of possible worlds or a 'truth set', which can be used to define logical properties like validity.

The FGCs of sentences, on the other hand, are genuine primitives in the sense that they are *not* inter-definable with other components of the framework. As a type of content, a FGC partitions worlds into those where it is true and those where it is not, but it is distinct from this corresponding 'truth set'. This allows for a two-level, composite analysis of the contents of sentences, whereby sentences are first assigned a FGC, and this in turn determines the 'truth set' or CGC of that sentence.

This is a generalization of neighborhood semantics. The *hyperintension* function H maps each sentence to a FGC, which the *intension* function I then maps to a CGC, and their composition satisfies classical operations on 'truth sets' such as $I(H(\neg\varphi)) = W \setminus I(H(\varphi))$. The knowledge operator is then interpreted along the lines of a modal operator from neighborhood semantics, i.e. there is a neighborhood function that has the job of assigning a set of 'known contents' to the designated agent. However, this is not a set of CGCs as in traditional neighborhood semantics, but a set of FGCs.

To extend this framework with a (normal) determinacy operator, we note that the access relation R of a Kripke model could always be replaced with a function $S(w) = \{x : wRx\}$ that outputs a *sphere* of alternatives to w. We assume that some truths are not determinate. This is represented by the existence of worlds in the sphere that disagree with each other. Determinate truths are those that hold the same in all worlds throughout this sphere.

A case of the IEC principle can then be captured by roughly the following modeling condition: whenever the FGC of φ is in the knowledge set of all worlds in the sphere of w, the FGC of ψ is in the knowledge set of some world in the sphere of w. This ensures that the agent determinately knows φ only if it is *indeterminate* that they fail to know ψ. Does this mean that this semantic method requires us to define the class of trivial consequences, once and for all? Fortunately not, as I will explain later.

Simple Semantics for Logics of Indeterminate Epistemic Closure

Let me first give the basic semantics for the following signature.

$$| \, p_i \, | \, \neg \, | \, \bot \, | \, \rightarrow \, | \, K \, | \, \Delta \, |$$

Definition 1 (Sphere Models) *A sphere model $M = \langle W, C, H, I, N, S \rangle$ is a six element structure such that...*

- $W \neq \emptyset$ *is a set of possible worlds*
- $C \neq \emptyset$ *is a set of fine-grained contents or FGCs*
- $H : Sent \rightarrow C$ *is a hyperintension function*
- $I : C \rightarrow \wp(W)$ *is an intension function*
- $N : W \rightarrow \wp(C)$ *is a neighborhood function, such that*
    - *If $c \in N(w)$, then $w \in I(c)$.*
- $S : W \rightarrow \wp(W)$ *is a sphere function, such that*
    - *Spheres are centered and jointly partition W.*

In addition, the coarse-grained content or CGC of φ in model M, given by the function $\llbracket \varphi \rrbracket^M = I(H(\varphi))$, must satisfy the following constraints.

- $\llbracket p \rrbracket^M = I(H(p))$ *for atomic p*
- $\llbracket \bot \rrbracket^M = \emptyset$
- $\llbracket \neg \varphi \rrbracket^M = W \setminus \llbracket \varphi \rrbracket^M$
- $\llbracket \varphi \rightarrow \psi \rrbracket^M = (W \setminus \llbracket \varphi \rrbracket^M) \cup \llbracket \psi \rrbracket^M$
- $\llbracket K \varphi \rrbracket^M = \{x \in W : H(\varphi) \in N(x)\}$
- $\llbracket \Delta \varphi \rrbracket^M = \{x \in W : S(x) \subseteq \llbracket \varphi \rrbracket^M\}$

The functions H and I can be naturally extended to sets of sentences and their contents, e.g. $H(\Gamma) = \{H(\psi) : \psi \in \Gamma\}$. For any given world $w \in W$ of a model, I refer to $N(w)$ as the *knowledge set* of that world. This is a framework for single-agent knowledge claims, but it can easily be extended to a multi-agent setting by parameterizing K and N to a set of agent names. Since this does not illuminate anything particularly interesting about the solution to the epistemic paradox, I leave aside such details.

The truth-conditions of a sentence are reflected in their CGCs, so we have that $K\varphi$ is true iff the FGC of φ is in the knowledge set according to w, and $\Delta\varphi$ is true iff φ is true throughout the sphere of w. The modeling condition on the neighborhood function, which coordinates known contents in N with intensions assigned by I, ensures factivity.

Here is a partial representation of a sphere model M to illustrate the main ideas of the semantics. In this model we have a four element Boolean algebra over a set of contents $C = \{c_1, c_2, c_3, c_4\}$ and we have a set of two worlds $W = \{w_1, w_2\}$ with 'universal' spheres $S(w_1) = W = S(w_2)$.

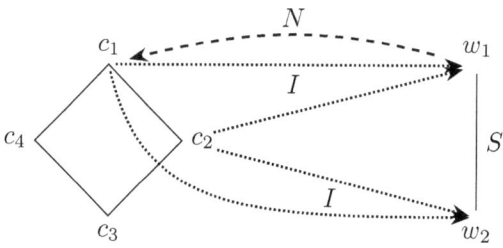

In a language of two atoms, we will let $H(p) = c_1$ and $H(q) = c_2$ and $H(\neg p) = c_3$ and $H(\neg q) = c_4$, with conjunctions and disjunctions assigned meets and joins of contents. Let $H(Kp) = c_2$ and $H(Kq) = c_3$ and $H(\Delta Kp) = c_3$. So, the top element is the FGC of p, the right is the FGC of q and Kp, and the bottom is the FGC of Kq and ΔKp.

The intension function maps $I(c_1) = W = I(c_2)$ as indicated in the diagram, but also $I(c_3) = \emptyset = I(c_4)$ which is not drawn. Finally, the neighborhood function maps $N(w_1) = \{c_1\}$ as indicated in the diagram, but also $N(w_2) = \emptyset$ which is not drawn. At world w_1, both p and q are true, but they are epistemically distinct because only p is known, furthermore the knowledge of p is not a determinate fact at w_1 because there is another world in its sphere w_2 where p is not known. Formally, we have:

- $[\![p]\!]^M = W$
- $[\![q]\!]^M = \{w_1\}$
- $[\![Kp]\!]^M \{w_1\}$
- $[\![Kq]\!]^M = \emptyset$
- $[\![\Delta Kq]\!]^M = \emptyset$

Simple Semantics for Logics of Indeterminate Epistemic Closure

In addition to reading them as truth-conditions, we also use the CGCs of sentences to define logical consequence as (local) truth-preservation.

Definition 2 ($\mathbb{C}$-Consequence) *The consequence relation of a class $\mathbb{C}$ of sphere models is defined by:* $\Gamma \vDash_\mathbb{C} \varphi$ *iff* $[\![\Gamma]\!]^M \subseteq [\![\varphi]\!]^M$ *in all* $M \in \mathbb{C}$.

This achieves many of our desiderata. On this semantics, knowledge is not closed under replacement of necessary equivalents. Since logical equivalence just consists in $[\![\varphi]\!]^M = [\![\psi]\!]^M$ in all models, we can model equivalent sentences as having distinct FGCs, so one can be known without the other. Notably, this 'fine-graining' is achieved without appeal to the metaphysics of truth-makers, situations, or impossible worlds. In addition, the knowledge and determinacy operators are factive and the determinacy operator is 'S5-like'. The modeling conditions on the sphere function ensure this because '... is in the sphere of...' is an equivalence relation.[7] These facts about sphere semantics are recorded below.

Remark 1 In the class $\mathbb{C}$ of all sphere models we have:

- $\varphi \leftrightarrow \psi \nvDash_\mathbb{C} K\varphi \leftrightarrow K\psi$ (K-**Hyperintensionality**)

- $\varphi \nvDash_\mathbb{C} \Delta\varphi$ (Δ-**Strength**)

- $K\varphi \vDash_\mathbb{C} \varphi$ (K-**Factivity**)

- $\Delta\varphi \vDash_\mathbb{C} \varphi$ (Δ-**Factivity**)

- $\neg\Delta\neg\varphi \vDash_\mathbb{C} \Delta\neg\Delta\neg\varphi$ (Δ-**Euclidean**)

- $\Delta(\varphi \to \psi) \vDash_\mathbb{C} \Delta\varphi \to \Delta\psi$ (Δ-**Distribution**)

- $\vDash_\mathbb{C} \varphi$ implies $\vDash \Delta\varphi$ (Δ-**Necessitation**)

What this does *not* yet achieve is the coordination of determinacy and knowledge operators that is required to validate the IEC principle. In the next section, I will explain how we can achieve this without actually defining the class of trivial consequences in advance, once and for all. The vagueness of triviality regulates this concept in a way that can be formally modeled.

---
[7]By the centering requirement, we always have $w \in S(w)$, and since a sphere is part of a partition, if $x \in S(w)$, then x and w simply have the same sphere $S(x) = S(w)$.

3 Respecting and regulating triviality

As a warm-up, I show how to validate an arbitrary case of IEC, assuming that we have identified a particular inference form as trivial. An inference form is a SET-FML pair. Suppose that the inference form $\langle \Gamma, \varphi \rangle$ is trivial. Call the following condition the Γ-φ *respect schema* (schematic over w).

(Γ-φ-RS) $H(\Gamma) \not\subseteq \bigcap\{N(x) : x \in S(w)\}$ or $H(\varphi) \in \bigcup\{N(x) : x \in S(w)\}$

If this holds at a given world w where it is a determinate truth that all the Γs are known, then it is indeterminate, at w, that knowledge of φ fails. This will be spelled out more carefully below. The upshot is that when this condition holds globally in a class of models, they validate the $\langle \Gamma, \varphi \rangle$ case of IEC.

Theorem 1 (The Γ-φ Respect Theorem) *For any given inference form $\langle \Gamma, \psi \rangle$, define a class $\mathbb{C}$ of sphere models as follows: $M \in \mathbb{C}$ iff (Γ-φ-RS) holds at all $w \in W$ of M. Then $\{\Delta K\psi : \psi \in \Gamma\} \vDash_{\mathbb{C}} \neg\Delta\neg K\varphi$.*

Proof. Let $M \in \mathbb{C}$ and $M = \langle W, C, H, I, N, S \rangle$. Let $w \in W$ and, for all $\psi \in \Gamma$, let $w \in [\![\Delta K\psi]\!]^M$. Then, for all $\psi \in \Gamma$ and all $x \in S(w)$, we have $x \in [\![K\psi]\!]^M$, equivalently, we have $H(\psi) \in N(x)$. So, for all $x \in S(w)$, we have $H(\Gamma) \subseteq N(x)$. In that case, $H(\Gamma) \subseteq \bigcap\{N(x) : x \in S(w)\}$ and so the condition (Γ-φ-RS) implies that $H(\varphi) \in \bigcup\{N(x) : x \in S(w)\}$ holds. It follows that $H(\varphi) \in N(x)$ for some $x \in S(w)$. So, $x \in [\![K\varphi]\!]^M$ for some $x \in S(w)$. Thus, $S(w) \not\subseteq [\![\neg K\varphi]\!]^M$ and so $w \in [\![\neg\Delta\neg K\varphi]\!]^M$. □

The question, then, is whether we can say anything more interesting about the logical role of triviality. Hypothetically, one could argue that we have an adequate solution to the paradox so long as we understand what it looks like to establish a modeling condition (Γ-φ-RS) for *whatever* inference forms $\langle \Gamma, \psi \rangle$ are trivial. Perhaps the question of which inference forms are genuinely trivial is only suited to informal philosophical debate.

This attitude, however, is unsatisfying if we want to develop the proof theory of IEC logics. If we leave 'trivial consequence' completely unsettled, it makes no sense to even try to axiomatize the relevant cases of the IEC principle (what are the relevant cases?). For this reason, I will attempt a more formal treatment of triviality. The trouble is that triviality is a vague concept. When applied to deductive inferences, 'trivial' connotes something like 'undeniably obvious'. There may be paradigm cases of this phenomenon, but there are also many contestable, borderline cases.

Simple Semantics for Logics of Indeterminate Epistemic Closure

At the outset, I suggested that primitive rules represent paradigm cases of triviality. Jago (2014a, p.11 and p.163) also endorses this claim and argues that it underwrites a *ranking* of candidate extensions of triviality. On the one hand, this implies that proof-theoretic factors constrain the concept of triviality. On the other hand, this claim does not take any official stand on the correct extension of the concept. I will show that this ranking approach does all of the work we need to axiomatize IEC logics.

In more detail, the ranking approach works as follows. Derivations are recursively defined. Longer derivations are formed by extending shorter derivations with a single application of one of the primitive rules. The *rank* of a consequence is the length of its shortest derivation. For each $n \in \mathbb{N}$, this gives us a precise candidate extension for the concept of triviality, by collecting all of the valid inference forms of lower ranks.[8]

The idea is that the proof theory of the base logic (without K and Δ) *grades* logical consequences. Semantics alone is not suited to this task because it only classifies inference forms as valid or invalid, without further distinctions. The gradation can be seen as a proxy for relative opacity: higher ranking consequences are those that are harder to recognize, so they are less fitting candidates for triviality.[9] What is interesting about this analysis is that it implies nothing about the correct extension of the concept of triviality, it merely tells us that there is a smooth transition from one candidate extension to another (which reflects the vagueness of triviality).

Nonetheless, this analysis provides a lot of interesting formal structure to work with. Since the primitive rules of the base logic establish the ranking, it can be implemented relative to different choices of base logic, so at an abstract level, this analysis is quite neutral about 'what is really trivial'. It does, however, imply that there are constraints on triviality once a base logic is fixed in the background. To assert that a high-ranking consequence is trivial implies that any lower-ranking consequence is also trivial.

For my purposes, the most useful aspect of this analysis is that it makes axiomatization possible. In the next section, I will implement the ranking approach relative to a Hilbert system and apply that ranking to give a full

---

[8]This strategy relates epistemic logic to the topic of proof complexity. Duc (1995) was the first person to mention this idea. Artemov and Kuznets (2014) use this idea to give a treatment of omniscience with operators $\Box_n$ meaning '... is known after n proof steps'.

[9]To say that one candidate is less fitting than another does not imply that it is unfit as such. All of these sets can be seen as genuine candidates in the sense that they could conceivably demarcate the extension of the concept of triviality.

semantics for a family of epistemic logics. This is best understood as one fully worked example of the analysis presented above.

4 A family of IEC logics

Consider a list of *potential* principles relating knowledge closure to certain valid inference forms of Classical Propositional Logic (CPL). We might ask ourselves: is knowledge of a tautology such as Excluded Middle indeterminate, or what about knowledge of a conclusion drawn by Modus Ponens? (we are asking about whether such consequences are trivial or not)

- $\vDash \neg\Delta\neg K(\varphi \vee \neg\varphi)$
- $\Delta K\neg\neg\varphi \vDash \neg\Delta\neg K\varphi$
- $\Delta K(\varphi \to \psi), \Delta K\varphi \vDash \neg\Delta\neg K\psi$
- $\Delta K(\varphi \vee \psi), \Delta K\neg\varphi \vDash \neg\Delta\neg K\psi$
- $\Delta K(\varphi \to \psi), \Delta K(\psi \to \chi) \vDash \neg\Delta\neg K(\varphi \to \chi)$

According to IEC theory, the validity of such inferences is always relativized to a candidate extension of triviality. Each principle holds relative to some candidates for triviality, but none hold absolutely.

We can make this precise. To do so, I will use a Hilbert system for CPL defined over the following signature (with other connectives definable).

$$\mid p_i \mid \neg \mid \bot \mid \to \mid$$

In order to implement the ranking approach, we want to think of derivations as the products of rules. Hilbert systems also usually have axioms, but we can capture this by stating a rule with no conditions, (R2) below, that effectively says that we can freely extend any derivation with any axiom. In the first instance, any rule-generated extension of the empty sequence of sentences counts as a derivation. Here are the axioms.

Definition 3 (Axioms of CPL) *The set of axioms AX_{CPL} of the background logic, CPL, is the set of all instances of the following schemata.*

(A1) $\varphi \to (\psi \to \varphi)$

(A2) $(\varphi \to (\psi \to \chi)) \to ((\varphi \to \psi) \to (\varphi \to \chi))$

(A3) $(\neg\varphi \to \neg\psi) \to (\psi \to \varphi)$

We use the axioms to define the primitive rules. The format of the rules, below, may look unusual, but they just re-write the usual defining conditions of a Hilbert derivation. In this formulation, primitive rules define admissible extensions of derivations (sequences of sentences).

Definition 4 (Primitive Rules of CPL) *A derivation from Γ in CPL is a finite (possibly empty) sequence of sentences $\psi_1, \ldots, \psi_n$. N.B. in the rules below, δ is used as a variable for such sequences.*

(R1) *If δ is a derivation from Γ and $\varphi \in \Gamma$, then δ, φ is a derivation from Γ.*

(R2) *If δ is a derivation from Γ and $\varphi \in AX_{CPL}$, then δ, φ is a derivation from Γ.*

(R3) *If δ is a derivation from Γ and there are members of this sequence of the form ψ and $\psi \to \varphi$, then δ, φ is a derivation from Γ.*

(R1) is the rule of assumptions, (R2) is the rule of free use of axioms, and of course (R3) is Modus Ponens. The set of all derivations from Γ is the smallest set of sequences closed under (R1)-(R3).

For the definition of derivability *per se*, we eliminate the empty sequence. Thus, the simplest inferences that are defined as derivable *per se* are one step derivations of assumptions or axioms.

Definition 5 (Derivability in CPL) *φ is derivable from Γ in CPL, written $\Gamma \vdash_{CPL} \varphi$, iff there is a non-empty derivation from Γ ending in φ, i.e. a sequence of sentences $\psi_1, \ldots, \psi_n$ that satisfies the criteria of Def. 4 such that $\psi_n = \varphi$. This sequence witnesses that $\Gamma \vdash_{CPL} \varphi$.*

This is the familiar derivability relation for CPL, re-written in a slightly unusual format. It is, however, worth making this explicit in order to clarify how this proof theory ranks the logical consequences of CPL. If the primitive rules (R1)-(R3) above, represent paradigm cases of triviality, then a specific ranking of sets of CPL consequences follows directly.

Definition 6 (Rank of CPL Consequences) *By the completeness theorem for CPL, if $\Gamma \vDash_{CPL} \varphi$, then there is a non-empty set of sequences of sentences $W(\Gamma, \varphi) = \{\delta : \delta \text{ witnesses that } \Gamma \vdash_{CPL} \varphi\}$. For each $\delta \in W(\Gamma, \varphi)$, let its length $\mathrm{len}(\delta)$ be equal to the number of sentences in the sequence. The rank of this consequence is $\#(\Gamma \vDash_{CPL} \varphi) = \min\{\mathrm{len}(\delta) : \delta \in W(\Gamma, \varphi)\}$.*

This ranking is what we care about. In a moment, I will point out how to use this ranking in an interesting way to generate IEC logics. First, here is a list of some examples to illustrate how various logical consequences of CPL are ranked by this system. Remember, this is meant to model relative opacity: higher ranks are less fitting candidates for triviality.

Remark 2 Ranking some logical consequences of CPL.

- $\#(\Gamma \vDash_{\text{CPL}} \varphi) = 1$ when $\varphi \in \Gamma$. **(Assumptions)**
- $\#(\Gamma \vDash_{\text{CPL}} \varphi \to (\psi \to \varphi)) = 1$ **(Axioms)**
- $\#(\Gamma, \varphi, \varphi \to \psi \vDash_{\text{CPL}} \psi) = 3$ **(Modus Ponens)**
- $\#(\Gamma, \neg\varphi \vDash_{\text{CPL}} \varphi \to \bot) = 5$ **(Negativity)**
- $\#(\Gamma, \varphi \to \psi, \psi \to \chi \vDash_{\text{CPL}} \varphi \to \chi) = 7$ **(Transitivity)**
- $\#(\Gamma, \varphi \to \psi \vDash_{\text{CPL}} \varphi \to (\psi \wedge \varphi)) = 9$ **(Pooling)**

Candidate extensions are sets of inference forms. The nth candidate extension for triviality collects all of SET-FML pairs corresponding to logical consequences of rank up to n. This candidate is called n-Triviality.

Definition 7 (n-Triviality) $\mathbb{T}_n = \{\langle \Gamma, \varphi \rangle : \#(\Gamma \vDash_{CPL} \varphi) \leq n\}$ *is the collection of all inference forms classified as 'trivial' by the nth candidate extension for triviality. I will write* $\mathbb{T}_n \langle \Gamma, \varphi \rangle$ *to mean* $\langle \Gamma, \varphi \rangle \in \mathbb{T}_n$.

Lemma 1 (Properties of n-Triviality) *The definition of $\mathbb{T}_n$ has all of the properties below, which leads Jago (2014b, p.1165) to refer to n-Triviality as '(a kind of) consequence' relation.*[10]

- *If* $\mathbb{T}_n \langle \Gamma, \varphi \rangle$, *then* $\Gamma \vDash_{CPL} \varphi$ *(Classicality)*
- $\mathbb{T}_n \langle \varphi, \varphi \rangle$ *(Reflexivity)*
- *If* $\mathbb{T}_n \langle \Gamma, \varphi \rangle$, *then* $\mathbb{T}_n \langle \Gamma \cup \Sigma, \varphi \rangle$ *(Monotonicity)*
- *If* $\mathbb{T}_n \langle \Gamma, \varphi \rangle$, *then* $\mathbb{T}_{n+1} \langle \Gamma, \varphi \rangle$ *(but not vice versa)* *(Heredity)*
- $\mathbb{T}_n \langle \Gamma, \varphi \rangle$ *and* $\mathbb{T}_n \langle \Gamma \cup \{\varphi\}, \psi \rangle$ *do* not *imply* $\mathbb{T}_n \langle \Gamma, \psi \rangle$ *('Cut' Fails)*

---

[10]This may seem odd since 'Cut' fails, but recent work on substructural logics makes this remark more reasonable (see, e.g., Cobreros, Egré, Ripley, & van Rooij, 2012).

Simple Semantics for Logics of Indeterminate Epistemic Closure

Proof. Straightforward. □

All of these properties are important, but Heredity in particular makes clear how this approach models the vagueness of triviality. It shows that, on this definition, there is a smooth transition from one candidate extension to another, which is why there are borderline cases of triviality.

Before returning to the full epistemic logic, we note one more useful fact. If $\mathbb{T}_n\langle \Gamma, \varphi \rangle$, then from Classicality, and the compactness theorem and deduction theorem, it follows that there is a finite set of assumptions $\{\psi_1, \ldots, \psi_m\} \subseteq \Gamma$ such that $\vdash_{CPL} \psi_1 \to (\ldots \to (\psi_m \to \varphi))$. Call this associated, derivable, nested implication sentence a *reductive implication*.

Definition 8 (Reductive Implications) *For each candidate extension $\mathbb{T}_n$, let RI_n be the set of all of its associated reductive implications.*

Note that an n-Trivial inference form can have more than one reductive implication. This redundancy is not important. We only need to know that there is at least one reductive implication for each n-Trivial inference.

We saw that if (Γ-φ-RS) holds at a world of a sphere model, then at that world, it is indeterminate whether knowledge is always closed under the inference from Γ to φ. The semantics for an IEC logic ought to *globally* respect every inference form that is considered to be trivial. I will use the precise candidates for triviality defined above to define a family of IEC logics, each of which respects (some candidate for) triviality.

So, returning to the full signature with operators K and Δ, and with sphere models defined as in §2, I will now focus on specific model classes. This defines not one unique logic, but a family of related logics.

Definition 9 (IEC Model Classes) *For each $n \in \mathbb{N}$, let $\mathbb{C}_n$ be the class of all sphere models that respect the candidate set $\mathbb{T}_n$ as follows:*

$$\mathbb{C}_n = \{M : \text{for all } \mathbb{T}_n\langle \Gamma, \varphi \rangle, (\Gamma\text{-}\varphi\text{-RS}) \text{ holds at all } w \in W \text{ of } M\}$$

For each candidate $\mathbb{T}_n$ above, we see that the IEC principles hold for *just those consequences* (considered as trivial) in the model class $\mathbb{C}_n$. The behavior of 'determinate knowledge', thus, depends purely on the structure of the notion of triviality. (cf. Jago 2014a, p.251, Theorem 8.3)

Corollary 1 *If $\mathbb{T}_n\langle \Gamma, \varphi \rangle$, then $\{\Delta K\psi : \psi \in \Gamma\} \vDash_{\mathbb{C}} \neg\Delta\neg K\varphi$.*

Proof. By an application of The Γ-ψ Respect Theorem. □

Since candidates for triviality increase in size $\mathbb{T}_n \subset \mathbb{T}_{n+1}$, as per the previous observation of Heredity, the corresponding model classes decrease in size $\mathbb{C}_{n+1} \subset \mathbb{C}_n$ and their consequence relations subsequently get stronger. In particular, the stronger logics in this family validate an increasing number of cases of the IEC principle. For example:

Remark 3 IEC principles for various $\mathbb{C}_n$-Consequence relations.

- $\Delta K(\varphi \to \psi), \Delta K\varphi \vDash_{\mathbb{C}_3} \neg\Delta\neg K\psi$ (3-MP)
- $\Delta K\neg\varphi \vDash_{\mathbb{C}_5} \neg\Delta\neg K(\varphi \to \bot)$ (5-Neg)
- $\Delta K(\varphi \to \psi), \Delta K(\psi \to \chi) \vDash_{\mathbb{C}_7} \neg\Delta\neg K(\varphi \to \chi)$ (7-Trans)
- $\Delta K(\varphi \to \psi) \vDash_{\mathbb{C}_9} \neg\Delta\neg K(\varphi \to (\psi \wedge \varphi))$ (9-Pool)

By way of illustration, consider the classically valid inference form of Transitivity and the 7-Trans result. According to these definitions, the narrowest candidates for triviality do not include Transitivity because it is not a paradigm case (not derivable by a single primitive rule). There are, however, more expansive candidate extensions of triviality that do consider the inference form Transivity to be trivial and for *any logic* that respects these candidates, the relevant IEC principle is valid (from $\mathbb{C}_7$ and above).

The family of $\mathbb{C}_n$-Consequence relations represent one fully worked example of the ranking approach to triviality. In this final section of the paper, I show how to axiomatize this family of logics.

5 Axiomatization

For each rank $n \in \mathbb{N}$, we can define a *logic of indeterminate closure* LIC_n as an extension of the classical Hilbert system for CPL. These have a shared, core set of axioms and rules, but they each have different axioms licensed by (A8) as the relevant versions of the IEC principle.

Definition 10 (Axioms of LIC_n) *For each $n \in \mathbb{N}$, the set of axioms AX_n is the set of all instances of the following schemata.*

(A1) $\varphi \to (\psi \to \varphi)$

(A2) $(\varphi \to (\psi \to \chi)) \to ((\varphi \to \psi) \to (\varphi \to \chi))$

(A3) $(\neg\varphi \to \neg\psi) \to (\psi \to \varphi)$

Simple Semantics for Logics of Indeterminate Epistemic Closure

(A4) $K\varphi \to \varphi$

(A5) $\Delta\varphi \to \varphi$

(A6) $\Delta\varphi \to \neg\Delta\neg\Delta\varphi$

(A7) $\Delta(\varphi \to \psi) \to (\Delta\varphi \to \Delta\psi)$

(A8) $\Delta K\psi_1 \to (\ldots \to (\Delta K\psi_m \to \neg\Delta\neg K\varphi))$
for all reductive implications $\psi_1 \to (\ldots \to (\psi_m \to \varphi)) \in RI_n$ [11]

Much as before, we use the axioms to define the primitive rules.

Definition 11 (Primitive Rules of LIC_n) *A derivation from Γ in LIC_n is a finite (possibly empty) sequence of sentences $\psi_1, \ldots, \psi_n$. N.B. in the rules below, δ is used as a variable for such sequences.*

(R1) *If δ is a derivation from Γ and $\varphi \in \Gamma$, then δ, φ is a derivation from Γ.*

(R2) *If δ is a derivation from Γ and $\varphi \in AX_n$, then δ, φ is a derivation from Γ.*

(R3) *If δ is a derivation from Γ and there are members of this sequence of the form ψ and $\psi \to \varphi$, then δ, φ is a derivation from Γ.*

(R4) *If δ is a derivation from Γ and there is a member of this sequence φ that does not depend on any $\psi \in \Gamma$, then $\delta, \Delta\varphi$ is a derivation from Γ.*

For the definition of derivability *per se*, we eliminate the empty sequence.

Definition 12 (Derivability in LIC_n) *φ is derivable from Γ in LIC_n, written $\Gamma \vdash_{\text{LIC}_n} \varphi$, iff there is a non-empty derivation from Γ ending in φ, i.e. a sequence of sentences $\psi_1, \ldots, \psi_n$ that satisfies the criteria of Def. 11 such that $\psi_n = \varphi$. This sequence witnesses that $\Gamma \vdash_{\text{LIC}_n} \varphi$.*

The completeness proof now follows by a canonical model construction. I will only sketch the interesting details. We first relativize the notions of a consistent and maximal set of sentences to each logic, LIC_n.

Definition 13 (n-Consistency) *$Con_n(\Gamma)$ iff for all sentences φ we have at least one of $\Gamma \nvdash_{\text{LIC}_n} \varphi$ or $\Gamma \nvdash_{\text{LIC}_n} \neg\varphi$*

---

[11] These are the sets RI_n from Definition 8, defined only over the signature of the base (classical) logic. Thanks to an anonymous referee for pointing out the potential ambiguity.

Definition 14 (n-Maximal-Consistency) $MCS_n(\Gamma)$ iff $Con_n(\Gamma)$ and it is not the case that $Con_n(\Gamma \cup \{\varphi\})$ for any $\varphi \notin \Gamma$

As usual, it follows that inconsistent sets can prove anything, and that $\Gamma \vdash_{LIC_n} \varphi$ iff it is *not* the case that $Con_n(\Gamma \cup \{\neg\varphi\})$. Maximal sets are deductively closed, and each logic supports Lindenbaum's Lemma.

Lemma 2 (Lindenbaum) *If $Con_n(\Gamma)$, then there is some sets of sentences Γ^* such that $\Gamma \subseteq \Gamma^*$ and $MCS_n(\Gamma^*)$.*

We define the proper canonical model and prove the Truth Lemma.

Definition 15 (The Proper Canonical Model of LIC_n) *Is a six element structure $\mathfrak{M}_n = \langle W, C, H, I, N, S \rangle$ with the following components.*

- $W_{\mathfrak{M}_n} = \{\Gamma : MCS_n(\Gamma)\}$

- $C_{\mathfrak{M}_n} = Sent$

- $H_{\mathfrak{M}_n}(\varphi) = \varphi$

- $I_{\mathfrak{M}_n}(\varphi) = \{\Gamma : \varphi \in \Gamma\}$

- $N_{\mathfrak{M}_n}(\Gamma) = \{\varphi : K\varphi \in \Gamma\}$

- $S_{\mathfrak{M}_n}(\Gamma) = \{\Sigma : \text{for all } \varphi \in Sent, \text{ if } \Delta\varphi \in \Gamma, \text{ then } \varphi \in \Sigma\}$

Lemma 3 (Truth) $\Gamma \in \llbracket \varphi \rrbracket^{\mathfrak{M}_n}$ iff $\varphi \in \Gamma$

Proof. This is quick: $\Gamma \in \llbracket \varphi \rrbracket^{\mathfrak{M}_n}$ iff $\Gamma \in I_{\mathfrak{M}_n}(H_{\mathfrak{M}_n}(\varphi))$ by the definition of CGCs iff $\varphi \in \Gamma$ by the definitions of $H_{\mathfrak{M}_n}$ and $I_{\mathfrak{M}_n}$. □

However, it still remains to establish that this simplistic structure *really is* a sphere model and that it belongs to the intended model class. For the first part, we need to see that the composition of $H_{\mathfrak{M}_n}$ and $I_{\mathfrak{M}_n}$ is well-behaved, i.e. that is satisfies all desired operations on 'truth sets'.

Lemma 4 ($\mathfrak{M}_n$ is a Sphere Model)

Proof. I present the illustrative cases of $\neg, K, \Delta$.

- $\Gamma \in \llbracket \neg\varphi \rrbracket^{\mathfrak{M}_n}$
  iff $\neg\varphi \in \Gamma$ by the Truth Lemma
  iff $\varphi \notin \Gamma$ by consistency
  iff $\Gamma \notin \llbracket \varphi \rrbracket^{\mathfrak{M}_n}$ by the Truth Lemma

Simple Semantics for Logics of Indeterminate Epistemic Closure

- $\Gamma \in \llbracket K\varphi \rrbracket^{\mathfrak{M}_n}$
  iff $K\varphi \in \Gamma$ by the Truth Lemma
  iff $\varphi \in N_{\mathfrak{M}_n}(\Gamma)$ by definition of $N_{\mathfrak{M}_n}$
  iff $H_{\mathfrak{M}_n}(\varphi) \in N_{\mathfrak{M}_n}(\Gamma)$ by definition of $H_{\mathfrak{M}_n}$

- $\Gamma \in \llbracket \Delta\varphi \rrbracket^{\mathfrak{M}_n}$
  iff $\Delta\varphi \in \Gamma$ by the Truth Lemma
  iff $\Sigma \in S_{\mathfrak{M}_n}(\Gamma)$ implies $\Sigma \in \llbracket \varphi \rrbracket^{\mathfrak{M}_n}$ by definition of $S_{\mathfrak{M}_n}$
  iff $S_{\mathfrak{M}_n}(\Gamma) \subseteq \llbracket \varphi \rrbracket^{\mathfrak{M}_n}$

□

Finally, we need to establish that this structure is in the right model class. That means: that the respect schema for all n-Trivial inference forms holds in all worlds (MCSs) of $\mathfrak{M}_n$. The following lemma will be useful.

Lemma 5 (Extension) *If $MCS_n(\Gamma)$ and $\Delta\varphi \notin \Gamma$, then in $\mathfrak{M}_n$ there is some $\Sigma \in S_{\mathfrak{M}_n}(\Gamma)$ such that $\varphi \notin \Sigma$. Contrapositively, we can infer that if all MCSs in the sphere of Γ in $\mathfrak{M}_n$ contain φ, then $\Delta\varphi \in \Gamma$.*

Proof. Let $MCS_n(\Gamma)$ and $\Delta\varphi \notin \Gamma$. I will first show that $\text{Con}_n(\Sigma')$ for the set $\Sigma' = \{\neg\varphi\} \cup \{\psi : \Delta\psi \in \Gamma\}$. Suppose not. Then there are finite $\psi_i \in \Sigma'$ such that $\vdash_{\text{LIC}_n} \psi_1 \to (\ldots \to (\psi_m \to \varphi))$. Since this sentence is derivable, it follows that $\vdash_{\text{LIC}_n} \Delta\psi_1 \to (\ldots \to (\Delta\psi_m \to \Delta\varphi))$ by the normality of Δ and so we have $\Gamma \vdash_{\text{LIC}_n} \Delta\varphi$. By hypothesis, however, we have $\Gamma \nvdash_{\text{LIC}_n} \Delta\varphi$, so indeed we have $\text{Con}_n(\Sigma')$ by reductio. We can apply Lindenbaum's Lemma to infer the existence of the target set: there is some Σ such that $\Sigma' \subseteq \Sigma$ and $MCS_n(\Sigma)$. By construction, if $\Delta\psi \in \Gamma$, then $\psi \in \Sigma$, so in the canonical model we have $\Sigma \in S_{\mathfrak{M}_n}(\Gamma)$ as desired, and $\neg\varphi \in \Sigma$. □

We show that the canonical model is in the right model class.

Lemma 6 (Rank) $\mathfrak{M}_n \in \mathbb{C}_n$

Proof. Let $\mathbb{T}_n \langle \Gamma, \varphi \rangle$. Note that there is at least one associated, reductive implication $\psi_1 \to (\ldots \to (\psi_m \to \varphi)) \in \text{RI}_n$ such that $\{\psi_1, \ldots, \psi_m\} \subseteq \Gamma$. Then by (A8) we have $\vdash_{\text{LIC}_n} \Delta K\psi_1 \to (\ldots \to (\Delta K\psi_m \to \neg\Delta\neg K\varphi))$, call this derivable sentence (*). Note that (*) is contained in all worlds of $\mathfrak{M}_n$. We can now show that (Γ-φ-RS) holds in all worlds. Suppose that $H_{\mathfrak{M}_n}(\Gamma) \subseteq \bigcap\{N_{\mathfrak{M}_n}(\Sigma) : \Sigma \in S_{\mathfrak{M}_n}(\Gamma)\}$. Let $\Pi \in S_{\mathfrak{M}_n}(\Gamma)$. It is then easy to work out that $\{K\psi_1, \ldots, K\psi_m\} \subseteq \Pi$ (with respect to the sentences $\{\psi_1,\ldots,\psi_m\} \subseteq \Gamma$ from above). Since this holds throughout $S_{\mathfrak{M}_n}$, it follows by the Extension Lemma that $\{\Delta K\psi_1,\ldots,\Delta K\psi_m\} \subseteq \Gamma$. Then since (*) is contained in Γ we have $\neg\Delta\neg K\varphi \in \Gamma$ by deductive closure and hence $\Delta\neg K\varphi \notin \Gamma$. It follows by the Extension Lemma that there is some $\Sigma \in S_{\mathfrak{M}_n}(\Gamma)$ such that $\neg K\varphi \notin \Sigma$ and hence $K\varphi \in \Sigma$. Thus, we have $H(\varphi) \in N_{\mathfrak{M}_n}\Sigma$ and hence $H_{\mathfrak{M}_n}(\varphi) \in \bigcup\{N_{\mathfrak{M}_n}(\Sigma) : \Sigma \in S_{\mathfrak{M}_n}(\Gamma)\}$. □

It follows that the logic LIC_n is sound and complete.

Theorem 2 (Adequacy) $\Gamma \vDash_{\mathbb{C}_n} \varphi$ iff $\Gamma \vdash_{LIC_n} \varphi$

Proof. Soundness is straightforward. For completeness, we reason that if $\Gamma \nvdash_{LIC_n} \varphi$, then $\text{Con}_n(\Gamma \cup \{\neg\varphi\})$, so there is a world (MCS) of the canonical model $\Sigma \in W_{\mathfrak{M}_n}$ with $\Gamma \cup \{\neg\varphi\} \subseteq \Sigma$, thus by the Truth Lemma, we have a member of the model class $\mathbb{C}_n$ which shows that $\Gamma \nvDash_{\mathbb{C}_n} \varphi$. □

6 Conclusion

IEC logics offer a formal solution to an epistemic paradox, by describing how trivial consequences are borderline cases of knowledge closure. I developed a simple semantics for these logics, without appeal to substantive metaphysics, and showed that such logics are axiomatizable.

In the process of showing these results, however, a number of question may have been raised. As emphasized in §3, the hard question for IEC theory is how to understand the concept of triviality. I described one way that proof-theory may constrain this concept.

The ranking approach to triviality provides formal structure, in the form of candidate extensions, that can be used to axiomatize IEC logics, but this is entirely determined by the choice of *primitive rules* of the base logic. So, from a philosophical point of view, there is more to say about the actual primitive rules that we 'really use' or that 'really give structure' to our concept of triviality. This is a difficult and important question, but one that lies beyond the scope of the present paper.

References

Artemov, S., & Kuznets, R. (2014). Logical omniscience as infeasibility. *Annals of Pure and Applied Logic, 165*(1), 6–25.

Cobreros, P., Egré, P., Ripley, D., & van Rooij, R. (2012). Tolerant, classical, strict. *Journal of Philosophical Logic*, *41*(2), 347–385.
Duc, H. N. (1995). Logical omniscience vs. logical ignorance: On a dilemma of epistemic logic. In C. Pinto-Ferreira & N. J. Mamede (Eds.), *Progress in Artificial Intelligence. EPIA 1995.* (pp. 237–248). Berlin, Heidelberg: Springer.
Fine, K. (1975). Vagueness, truth and logic. *Synthese*, *30*(4), 265–300.
Hawke, P., Özgün, A., & Berto, F. (2020). The fundamental problem of logical omniscience. *Journal of Philosophical Logic*, *49*(4), 727–766.
Hintikka, J. (1962). *Knowledge and Belief: an Introduction to the Logic of the Two Notions.* Ithaca: Cornell University Press.
Jago, M. (2006). Hintikka and Cresswell on logical omniscience. *Logic and Logical Philosophy*, *15*(4), 325–354.
Jago, M. (2014a). *The Impossible: An Essay on Hyperintensionality.* Oxford: Oxford University Press.
Jago, M. (2014b). The problem of rational knowledge. *Erkenntnis*, *79*(suppl. 6), 1151–1168.
Sedlár, I. (2021). Hyperintensional logics for everyone. *Synthese*, *198*(2), 933–956.
Yablo, S. (2014). *Aboutness.* Princeton: Princeton University Press.

Colin R. Caret
Utrecht University, Department of Philosophy and Religious Studies
Netherlands
E-mail: `colin.caret@protonmail.com`

Towards Tractable Approximations to Many-Valued Logics: The Case of First Degree Entailment

MARCELLO D'AGOSTINO[1] AND ALEJANDRO SOLARES-ROJAS[2]

Abstract: **FDE** is a logic that captures relevant entailment between implication-free formulae and admits of an intuitive informational interpretation as a 4-valued logic in which "a computer *should* think". However, the logic is co-NP complete, and so an idealized model of how an agent *can* think. We address this issue by shifting to signed formulae where the signs express imprecise values associated with two distinct bipartitions of the set of standard 4 values. Thus, we present a proof system which consists of linear operational rules and only two branching structural rules, the latter expressing a generalized rule of bivalence. This system naturally leads to defining an infinite hierarchy of tractable depth-bounded approximations to **FDE**. Namely, approximations in which the number of nested applications of the two branching rules is bounded.

Keywords: **FDE**, tractability, natural deduction, tableaux

1 Introduction

Many interesting propositional logics are likely to be computationally intractable. For instance, Classical Propositional Logic (**CPL**) and First-Degree Entailment (**FDE**; Anderson & Belnap, 1962; Belnap, 1977a, 1977b; Dunn, 1976) are both co-NP complete (Arieli & Denecker, 2003; Cook, 1971; Urquhart, 1990). Thus, we cannot expect a real agent, no matter whether human or artificial, to be always able to recognize in practice that a certain conclusion follows from a given set of assumptions. This is a source of

---

[1] We thank two anonymous reviewers, Miguel Pérez-Gaspar and Pietro Casati for helpful comments. This research was funded by the Department of Philosophy "Piero Martinetti" of the University of Milan under the Project "Departments of Excellence 2018-2022" awarded by the Ministry of Education, University and Research (MIUR). The first author was also partly supported by MIUR under the PRIN 2017 project (n. 20173YP4N3).

[2] The second author was partly supported by the Israel Science Foundation (grant 550/19).

major difficulties in research areas that are in need of less idealized, yet theoretically principled, models of logical agents with bounded cognitive and computational resources. The "depth-bounded approach" to **CPL** (e.g., D'Agostino, 2015; D'Agostino, Finger, & Gabbay, 2013; D'Agostino & Floridi, 2009) provides an account of how this logic can be approximated in practice by realistic agents in two moves: i) by providing a semantic and proof-theoretic characterization of a tractable 0-depth approximation, and ii) by defining an infinite hierarchy of tractable k-depth approximations, which can be naturally related to a hierarchy of realistic, resource-bounded agents, and admits of an elegant proof-theoretic characterization.

A key idea underlying the "depth-bounded approach" to **CPL** is that the meaning of a logical operator is specified solely in terms of the information that is actually possessed by an agent, i.e. information practically accessible to her and with which she can operate. This kind of information is called *actual*, and we use the verb "to hold" as synonymous with "to actually possess". The semantics is ultimately based on intuitive, albeit non-deterministic, 3-valued tables that were first put forward by Quine (1973) to capture the "primitive meaning of the logical constants". The values have a natural informational interpretation ("accept", "reject", "abstain"). The proof-theoretic characterization given in (D'Agostino, 2015; D'Agostino et al., 2013) is based on introduction and elimination (intelim) rules that, unlike those of Gentzen-style natural deduction, involve no "discharge" of hypotheses. The *0-depth approximation* consists of the consequence relation associated with the intelim rules only, is computationally easy (tractable) and corresponds to Quine's non-deterministic semantics. The *depth* of **CPL** inferences is measured in terms of the maximum number of nested applications of a single *branching* rule, which is a Classical Dilemma rule called *PB* ("Principle of Bivalence"). *PB* governs the manipulation of *virtual* information, i.e., hypothetical information that an agent does not hold, but she temporarily *assumes as if* she held it. Intuitively, the more times such virtual information needs to be invoked via *PB*, the harder the corresponding inference is for any agent who is able to perform at least 0-depth inferences, both from the computational and the cognitive point of view. In essence, each k-depth logic corresponds to a limited capability of manipulating virtual information.

The depth-bounded approach to **CPL**, as remarked in (D'Agostino, 2015), is the first stage of a more general research program that aims to define similar approximations to first-order logic and to a variety of non-classical logics. A preliminary step towards the first order case can be found in (D'Agostino, Larese, & Modgil, 2021). In this paper we show how the depth-bounded approach can be naturally extended to a useful many-valued system such as **FDE**. This provides a case study for extending the depth-bounded framework to a variety of finite-valued logics, in the spirit of (Caleiro, Marcos, & Volpe, 2015; Carnielli, 1987; Hähnle, 1999).

2 Belnap's semantics and the need for imprecise values

FDE arose out of the study of relevance logics.[3] Based on work of Dunn (1976), and an observation by Smiley (in correspondence), Belnap (1977a, 1977b) gave an interesting semantic characterization of **FDE** in terms of a 4-valued logic, and pointed out its usefulness as the logic in which "a computer should think". This characterization has become not only the standard semantics of **FDE**, but also its standard presentation. It is motivated from the use of deductive reasoning as a basic tool in the area of "intelligent" database management or question-answering systems. Databases have a great propensity to be incomplete and become inconsistent: what is stored in a database is usually obtained from different sources which may provide only partial information and may well conflict with each other. For a matrix to characterize a logic adequate for making deductions with information that might be both inconsistent and partial, at least 4 different values are needed (see Arieli, Avron, & Zamansky, 2018). An elegant 4-valued matrix is precisely Belnap-Dunn's.

In what follows we assume a standard propositional language $\mathcal{L}$ with $\wedge$, $\vee$ and $\neg$ as logical operators. We use $p, q, r, ...$, possibly with subscripts, as metalinguistic variables for atomic formulae; $A, B, C, ...$, possibly with subscripts, for arbitrary formulae; and $\Gamma, \Delta, \Lambda, ...$, possibly with subscripts, to vary over sets of formulae. Let $F(\mathcal{L})$ and $At(\mathcal{L})$ respectively be the set of well-formed formulae and atomic formulae of $\mathcal{L}$. Moreover, we define the *degree* of a formula A as the number of occurrences of connectives in A. The set of truth-values is $\{\mathbf{t}, \mathbf{f}, \mathbf{b}, \mathbf{n}\}$ and is denoted by **4**. These values are interpreted as four possible ways in which an atom p can belong to the present state of information of a computer's database, which in turn is fed by a set of equally "reliable" sources: $\mathbf{t}$ means that the computer is told that p is true by some source, without being told that p is false by any source; $\mathbf{f}$ means that the computer is told that p is false but never told that p is true; $\mathbf{b}$ means that the computer is told that p is true by some source and that p is false by some other source (or by the same source in different times); $\mathbf{n}$

---

[3]For a survey on **FDE** see (Omori & Wansing, 2017).

means that the computer is told nothing about the value of p. In essence, each value represents a subset of $\{true, false\}$ (Dunn, 1976). These four values form two distinct lattices, depending on whether we consider the partial information ordering induced by set-inclusion (*approximation* lattice) or the partial ordering based on "closeness to the truth" (*logical* lattice). The information ordering is the one according to which the epistemic state of the computer concerning an atom can evolve over time. As Belnap points out:

> When an atomic formula is entered into the computer as either affirmed or denied, the computer modifies its current set-up by adding a "told True" or "told False" according as the formula was affirmed or denied; it does not subtract any information it already has [...] In other words, if p is affirmed, it marks p with **t** if p were previously marked with **n**, with **b** if p were previously marked with **f**; and of course leaves things alone if p was already marked either **t** or **b** (Belnap, 1977a, p. 12).

A *set-up* is simply an assignment to each of the atoms of exactly one of the values in **4**. The values of complex formulae are obtained by means of considerations related to "Scott's thesis" about approximation lattices (Belnap, 1977a), resulting in the truth-tables in Table 1. Using these truth-tables, every set-up can be extended to a valuation function $v : F(\mathcal{L}) \longrightarrow \mathbf{4}$ in the usual inductive way. We call this function a **4**-*valuation*. It establishes how the computer is to answer questions about complex formulae based on a set-up. While answering questions on the basis of a given epistemic set up is computationally easy, we do not have a logic yet. As Belnap puts it, we "want some rules for the computer to use in generating what it implicitly knows from what it explicitly knows", i.e., we need a logic for the computer to reason.[4] This is achieved by turning Belnap-Dunn's matrix into a valuation system where the set $\mathcal{V}$ of values is equal to **4**, and the set $\mathcal{D}$ of designated values is equal to $\{\mathbf{t}, \mathbf{b}\}$. (Warning: do not confuse the values in **4** with *true* and *false*. The latter are *local* values referring to the information coming from a source, the former are *global* values, summarizing the epistemic state of the computer with respect to all the sources.) The consequence relation is then defined as follows:

Definition 1 $\Gamma \vDash A$ *iff for every* **4**-*valuation* v, *if* $v(B) \in \{\mathbf{t}, \mathbf{b}\}$ *for all* $B \in \Gamma$, *then* $v(A) \in \{\mathbf{t}, \mathbf{b}\}$.

---

[4]As one anonymous reviewer pointed out to us, there is a tension between a justification in terms of information and the propositional attitude of knowledge in Belnap's seminal papers (Belnap, 1977a, 1977b), which is addressed in (Wansing & Belnap, 2010).

Tractable Depth-Bounded Approximations to FDE

∨	t	f	b	n		∧	t	f	b	n		¬	
t	t	t	t	t		t	t	f	b	n		t	f
f	t	f	b	n		f	f	f	f	f		f	t
b	t	b	b	t		b	b	f	b	f		b	b
n	t	n	t	n		n	n	f	f	n		n	n

Table 1: **FDE**-tables

As is well-known, the relation ⊨ above is a (finitary) Tarskian consequence relation. That is, it satisfies the following conditions:

Reflexivity: If $A \in \Gamma$, then $\Gamma \vDash A$.

Monotonicity: If $\Gamma \vDash A$, then $\Gamma \cup \Delta \vDash A$.

Cut: If $\Gamma \vDash A$ and $\Gamma \cup \{A\} \vDash B$, then $\Gamma \vDash B$.

Further, for the unrestricted language allowing arbitrary formulae involving ∧, ∨ and ¬, the decision problem for this consequence relation is co-NP complete (see, Arieli & Denecker, 2003; Urquhart, 1990), which brings us to the need for tractable approximations. In the next section we shall present a sort of natural deduction system for **FDE** based on two key observations.

First, as implicit in the quotation from Belnap above, the values in 4, except for **b**, cannot be taken as *stable*. An epistemic set up is just a snapshot of an epistemic state that evolves over time. If we want to consider the truth-values **t**, **f**, **n** as stable we need to assume complete information about the set of sources Ω. Namely, while the meaning of **b** is "*there is* at least a source assenting to p and at least a source dissenting from p" (which is information empirically accessible to x in the sense that x may *hold* this information without a complete knowledge of Ω), the meaning of **t**, **f** and **n** involves information of the kind "*there is no* source such that...", and so requires complete information about the sources in Ω, which may not be empirically accessible to x at any given time. What if the agent does not have such a complete knowledge about the sources? For instance, the agent may well be receiving information from an "open" set of sources as they become accessible (even if the information coming from each single source is assumed to be robust). In such a case, the possibility for an agent to come across a source falsifying "there is no source such that..." is always open. Thus, despite their informational nature, three of the values in 4 are *information-transcendent* when interpreted as timeless. They refer to an objective state of affairs concerning the domain of all sources, that may well be inaccessible to the computer at any given time. This motivates the need for *a stable imprecise value* such as "t *or* b", which is implicit in the choice of the set of designated values by Belnap. Inspired by (D'Agostino, 1990) and (Avron, 2003; Fitting, 1991, 1994), we shall address this question by shifting to *signed formulae*, where the signs express such imprecise values associated with two distinct bipartitions of 4.

A *second* key observation is that, as suggested by Belnap (1977a, 1977b), there is no reason to assume that an agent is "told" about the values of *atoms only*. As we shift from objective truth and falsity to informational truth and falsity, this is a highly unrealistic restriction. In most practical contexts we may be told that a certain disjunction is true without being told which of the two disjuncts is the true one, or that a certain conjunction is false without being told which of the two conjuncts is the false one. As a simple example of the former situation, take the information that Alice and Bob are siblings (either they have the same mother or they have the same father); for the latter, take the information that Alice and Bob are not siblings, i.e., for any individual x, the conjunction "x is a parent of Bob and x is a parent of Alice" must be false, which amounts to saying that either the first or the second conjunct is false, without necessarily knowing which. In the context of **CPL**, these considerations naturally lead to a non-deterministic 3-valued semantics which was anticipated by Quine. (See D'Agostino (2014) for further references and a discussion that includes an interesting quotation from Michael Dummett to the effect that in non-mathematical contexts our information may well be *irremediably disjunctive* in nature.)

These two observations prompt us to propose a proof-theoretic approach to depth-bounded reasoning in **FDE** that is similar to the one taken in (D'Agostino, 2015; D'Agostino et al., 2013; D'Agostino & Floridi, 2009) for **CPL**. Before addressing this issue, however, we shall provide in the next section a proof-theoretic characterization of depth-unbounded reasoning in **FDE** that will pave the way for defining its tractable approximations.

3 Intelim deduction in FDE

Signed formulae (S-formulae, for short) are expressions of the form T A, F A, T* A and F* A. Denoting an agent with x and a 4-valuation with v, their intended interpretation is respectively as follows: "x holds that A is at least true" (expressing that $v(A) \in \{\mathbf{t}, \mathbf{b}\}$); "$x$ holds that A is non-true" ($v(A) \in \{\mathbf{f}, \mathbf{n}\}$); "$x$ holds that A is non-false" ($v(A) \in \{\mathbf{t}, \mathbf{n}\}$); "$x$ holds that A is at least false" ($v(A) \in \{\mathbf{f}, \mathbf{b}\}$).[5] Crucially, S-formulae of the form T A or F* A express information that x may hold even without a complete knowledge of the set of sources Ω. However, this is not the case of the other two types of S-formulae which involve complete knowledge of Ω and so can only be assumed hypothetically. Now, we say that the *conjugate* of T A is F A and vice versa, and that the conjugate of T* A is F* A and vice versa. Besides, we write T Γ for $\{T A \mid A \in \Gamma\}$. Moreover, we use $\varphi, \psi, \theta, ...$, possibly with subscripts, as variables ranging over S-formulae; and $X, Y, Z, ...$, possibly with subscripts, as variables ranging over sets of S-formulae. Further, let us use $\bar{\varphi}$ to denote the conjugate of φ. Finally, we say that the *unsigned part* of an S-formula is the unsigned formula that results from it by removing its sign. Given an S-formula φ, we denote by φ^u the unsigned part of φ and by X^u the set $\{\varphi^u \mid \varphi \in X\}$. Note also that, for the reasons explained in the previous section, an agent may hold the information that T $A \lor B$, but neither the information that T A nor that T B. Similarly, she may hold the information that F* $A \land B$, but neither the information that F* A nor that F* B.

We identify the basic (*0-depth*) logic of our hierarchy of approximations with the inferences that an agent can draw without making hypotheses about the "objective" state of affairs concerning the whole of Ω. In other words, without making hypothetical assumptions that go beyond the information that she holds. We shall show that a natural proof-theoretic characterization of this basic logic is obtained by means of the set of introduction and elimination (*intelim*) rules respectively displayed in Tables 2 and 3. The analogous 0-depth system for **CPL** in (D'Agostino, 2015; D'Agostino et al., 2013) is characterized by the intelim rules obtained by removing all the starred signs, replacing them with the unstarred signs T and F, interpreted as "only true" and "only false", and eliminating duplicates. Note that the characterization

---

[5] Similar approaches to **FDE** are given in (Blasio, 2015, 2017) and (Shramko & Wansing, 2005), but they are extended along very different lines and used for very different purposes. Particularly, in those approaches there is no attempt to provide tractable approximations. We thank Luis Estrada-González for having pointed us at the latter approach.

$$\frac{\mathsf{F}\,A}{\mathsf{F}\,A \wedge B} \qquad \frac{\mathsf{F}\,B}{\mathsf{F}\,A \wedge B} \qquad \frac{\mathsf{F}^*\,A}{\mathsf{F}^*\,A \wedge B} \qquad \frac{\mathsf{F}^*\,B}{\mathsf{F}^*\,A \wedge B}$$

$$\frac{\mathsf{T}\,A}{\mathsf{T}\,A \vee B} \qquad \frac{\mathsf{T}\,B}{\mathsf{T}\,A \vee B} \qquad \frac{\mathsf{T}^*\,A}{\mathsf{T}^*\,A \vee B} \qquad \frac{\mathsf{T}^*\,B}{\mathsf{T}^*\,A \vee B}$$

$$\frac{\mathsf{T}\,A \quad \mathsf{T}\,B}{\mathsf{T}\,A \wedge B} \qquad \frac{\mathsf{F}\,A \quad \mathsf{F}\,B}{\mathsf{F}\,A \vee B} \qquad \frac{\mathsf{T}^*\,A \quad \mathsf{T}^*\,B}{\mathsf{T}^*\,A \wedge B} \qquad \frac{\mathsf{F}^*\,A \quad \mathsf{F}^*\,B}{\mathsf{F}^*\,A \vee B}$$

$$\frac{\mathsf{T}\,A}{\mathsf{F}^*\,\neg A} \qquad \frac{\mathsf{F}\,A}{\mathsf{T}^*\,\neg A} \qquad \frac{\mathsf{T}^*\,A}{\mathsf{F}\,\neg A} \qquad \frac{\mathsf{F}^*\,A}{\mathsf{T}\,\neg A}$$

Table 2: Introduction rules for the standard **FDE** connectives

of the basic logic bears some resemblance with natural deduction, but does not have discharge rules, since no hypothetical reasoning is involved. In the elimination rules, we shall refer to the premise containing the connective that is to be eliminated as *major* and to the other premise as *minor*. In turn, given that the intelim rules have all a linear format, their application generates *intelim sequences*. Namely, finite sequences $(\varphi_1, ..., \varphi_n)$ s.t., for every $i = 0, ..., n$, either φ_i is an assumption or it is the conclusion of the application of an intelim rule to preceding S-formulae.

The intelim rules are all sound, but not complete for full **FDE**. Indeed, as we shall show, these rules just characterize the basic logic in the hierarchy. Completeness is obtained by adding only two branching structural rules, according to which we are allowed to: (i) append both T A and F A as sibling nodes to the last element of any intelim sequence; (ii) append both T* A and F* A in a similar way. Their intuitive meaning is that one of the two cases must obtain considering the whole of Ω even if the agent has no actual information about which is the case. In this sense, we call the information expressed by each conjugate S-formula *virtual* information; i.e., hypothetical information that the agent does not hold, but she temporarily *assumes as if* she held it. We respectively call these branching rules PB and PB^* as they are closely related to a *generalized* Principle of Bivalence:[6]

---

[6]Generalizations of the rule of bivalence have been fruitfully used in the context of many-valued and substructural logics (see Caleiro et al., 2015; D'Agostino, Gabbay, & Broda, 1999; Hähnle, 1999).

Tractable Depth-Bounded Approximations to **FDE**

$$\frac{F\,A\wedge B \quad T\,A}{F\,B} \qquad \frac{F\,A\wedge B \quad T\,B}{F\,A} \qquad \frac{F^*\,A\wedge B \quad T^*\,A}{F^*\,B} \qquad \frac{F^*\,A\wedge B \quad T^*\,B}{F^*\,A}$$

$$\frac{T\,A\wedge B}{T\,A} \qquad \frac{T\,A\wedge B}{T\,B} \qquad \frac{T^*\,A\wedge B}{T^*\,A} \qquad \frac{T^*\,A\wedge B}{T^*\,B}$$

$$\frac{T\,A\vee B \quad F\,A}{T\,B} \qquad \frac{T\,A\vee B \quad F\,B}{T\,A} \qquad \frac{T^*\,A\vee B \quad F^*\,A}{T^*\,B} \qquad \frac{T^*\,A\vee B \quad F^*\,B}{T^*\,A}$$

$$\frac{F\,A\vee B}{F\,A} \qquad \frac{F\,A\vee B}{F\,B} \qquad \frac{F^*\,A\vee B}{F^*\,A} \qquad \frac{F^*\,A\vee B}{F^*\,B}$$

$$\frac{T\,\neg A}{F^*\,A} \qquad \frac{F\,\neg A}{T^*\,A} \qquad \frac{T^*\,\neg A}{F\,A} \qquad \frac{F^*\,\neg A}{T\,A}$$

Table 3: Elimination rules for the standard **FDE** connectives

$$\mathsf{T}\,A \mid \mathsf{F}\,A \qquad \mathsf{T^*}\,A \mid \mathsf{F^*}\,A$$

For **CPL** only the first rule, with T and F interpreted as "only true" and "only false", makes sense and is sufficient for completeness. With the addition of these rules, deductions are represented by downward-growing trees, which brings the method somewhat closer to tableaux.[7] Each application of PB or PB^* stands for the introduction of virtual information about the imprecise value of a formula A, which we shall respectively call the PB-*formula* or PB^*-*formula*. Note once again that, whereas signed formulae of the form T A and F* A are empirically obtainable, signed formulae of the form T* A and F A express hypotheses introduced by applications of PB or PB^*. In turn, the S-formulae T A, F A, T* A and F* A appended via those branching rules will be all called *virtual assumptions*. Now, PB and PB^* are essentially cut rules that may introduce formulae of arbitrary degree. However, as we will show in Lemma 2, their application can be restricted so as to satisfy the subformula property. Moreover, from our informational viewpoint, the

---

[7] Well-known (Smullyan-style) tableaux for **FDE** were introduced by Priest (2001). Space prevent us to include a fair comparison of our investigation in this paper with related work; we shall include it in a subsequent paper.

main conceptual advantage of this proof-theoretic characterization consists in that it clearly separates the *intelim rules* that fix the meaning of the connectives in terms of the information that an agent holds from the two *structural rules* that introduce virtual information (PB and PB^*).

Intuitively, the more virtual information needs to be invoked via PB or PB^*, the harder the inference is for the agent, both from the computational and the cognitive viewpoint. In this sense, the nested applications of PB and PB^* provide a sensible measure of inferential *depth*. This naturally leads to defining an infinite hierarchy of tractable depth-bounded approximations to **FDE** in terms of the maximum number of nested applications of PB and PB^* that are allowed. Before giving definitions and results, we remark that (i) unlike the branching rules of Smullyan-style tableaux, our branching rules are *structural* in that they do not involve any specific logical operator; (ii) the elimination rules, together with the branching rules, were early introduced in (D'Agostino, 1990) as constituting a refutation method for full **FDE** called RE_{fde}. So, the completeness of RE_{fde} trivially implies the completeness of the system presented in this paper. However, our intelim method can be used as a direct-proof method as well as a refutation method, and leads to more powerful approximations. A direct completeness proof can also be given based on the semantics, which implies the subformula property. In this paper we choose to prove a more general version of the subformula property by means of proof transformations.

Definition 2 *Let $X = \{\varphi_1, ..., \varphi_m\}$. Then $\mathcal{T}$ is an* intelim tree *for X if there is a finite sequence $(\mathcal{T}_1, \mathcal{T}_2, ..., \mathcal{T}_n)$ s.t. $\mathcal{T}_1$ is a one-branch tree consisting of any sequence of the formulae in X, $\mathcal{T}_n = \mathcal{T}$, and for each $i < n$, $\mathcal{T}_{i+1}$ results from $\mathcal{T}_i$ by an application of an intelim rule to preceding S-formulae in the same branch, or by an application of PB or PB^*. A branch of an intelim tree is* closed *if it contains an S-formula φ and its conjugate $\bar{\varphi}$; otherwise, it is* open. *An intelim tree is said to be* closed *when all its branches are closed; otherwise, it is* open. *An* intelim proof *of φ from X is an intelim tree $\mathcal{T}$ for X s.t. φ occurs in all open branches of $\mathcal{T}$. An* intelim refutation *of X is a closed intelim tree $\mathcal{T}$ for X.*

Note that every refutation of X is, simultaneously, a proof of φ from X, for every φ. This is, of course, a kind of explosivity; but it regards *signed* formulae, and it is compatible with the non-explosivity regarding formulae in **FDE**. The reason of that compatibility is that a set consisting of S-formulae all of the form T A cannot lead to explosion because there cannot be an intelim refutation of such a set:

Proposition 1 *Any intelim tree for a set* T Γ *has at least a branch containing only S-formulae of the form* T A *or* F$^*A$.

Proof. By an easy induction on the number of nodes of the intelim tree. □

4 Subformula property

A proof has the *subformula property* if every formula in it is a subformula either of the assumptions or of the conclusion. In the case of refutations only subformulae of the assumptions can occur.

Definition 3 *An occurrence of an S-formula φ in an intelim tree $\mathcal{T}$ is: (i) a* detour *if φ is both the conclusion of an introduction and the major premise of an elimination; (ii)* idle *if it is not the terminal node of its branch, it is not used as premise of some application of an intelim rule, and it is not the conjugate of some S-formula occurring in the same branch.*

Definition 4 *Let $\mathcal{T}$ be an intelim proof of φ from X (an intelim refutation of X). $\mathcal{T}$ is* non-redundant *if it satisfies the following conditions: (i) it contains no idle occurrences of S-formulae; (ii) none of its branches contains more than one occurrence of the same S-formula; (iii) none of its branches properly includes a closed path.*

Lemma 1 *If an intelim proof or refutation $\mathcal{T}$ is non-redundant, then it contains no detours.*

Proof. By the definitions above and inspection of the intelim rules. Every detour makes the tree redundant. □

We observe that turning an intelim proof or refutation $\mathcal{T}$ into a non-redundant one is computationally easy. Now, let us denote by sub(Δ) the closure under subformulae of the set Δ of formulae. Given a proof $\mathcal{T}$ of φ from X (a refutation of X) we say that an application of PB or PB^* in $\mathcal{T}$ is *analytic* if its PB-formula or PB^*-formula is in sub($X^u \cup \{\varphi^u\}$) (sub(X^u)). We also say that an intelim tree is *analytic* if all the applications of PB and PB^* in it are analytic.

Lemma 2 *Every intelim proof $\mathcal{T}$ of φ from X (refutation of X) can be transformed into a proof $\mathcal{T}'$ of φ from X (refutation $\mathcal{T}'$ of X) s.t. every application of PB and PB^* in $\mathcal{T}$ is analytic.*

Proof. We use the notation $\overset{\mathcal{T}}{\varphi}$ to denote either the empty intelim tree or a non-empty intelim tree s.t. φ is one of its terminal nodes. The proof is by lexicographic induction on $\langle \gamma(\mathcal{T}), \kappa(\mathcal{T}) \rangle$, where $\gamma(\mathcal{T})$ denotes the maximum degree of a PB-formula or a PB^*-formula in $\mathcal{T}$ that is not analytic, and $\kappa(\mathcal{T})$ denotes the number of occurrences of such non-analytic PB-formulae or PB^*-formulae of maximal degree. Let $\gamma(\mathcal{T}) = m > 0$ and let A be a PB-formula or a PB^*-formula of degree m. There are several cases depending on the logical form of A and on whether A is a PB-formula or a PB^*-formula. We sketch only the case where $A = B \wedge C$ and A is a PB^*-formula; the other cases being similar. So, $\mathcal{T}$ has the following form:

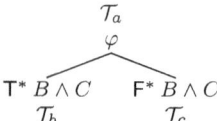

where $\mathcal{T}_b$ and $\mathcal{T}_c$ are intelim trees s.t. each of their open branches contains φ, or are both closed intelim trees. Let $\mathcal{T}'$ be the following intelim tree:

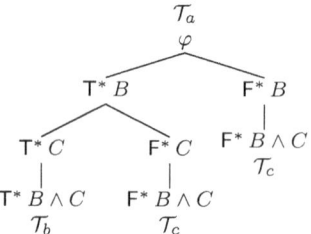

Clearly, $\mathcal{T}'$ is a proof of φ from X (refutation of X). Moreover, either $\gamma(\mathcal{T}') < \gamma(\mathcal{T})$, or $\gamma(\mathcal{T}') = \gamma(\mathcal{T})$ and $\kappa(\mathcal{T}') < \kappa(\mathcal{T})$. □

Given Lemma 1 and Lemma 2, it is not difficult to show that:

Theorem 1 (Subformula property) *If $\mathcal{T}$ is a proof of φ from X (a refutation of X), $\mathcal{T}$ can be transformed into a proof (refutation) $\mathcal{T}'$ of φ from X (of X) with the subformula property.*

5 Depth-bounded approximations to FDE

Definition 5 *The* depth *of an intelim tree $\mathcal{T}$ is the maximum number of virtual assumptions occurring in a branch of $\mathcal{T}$. An intelim tree $\mathcal{T}$ is a*

k-depth intelim proof of φ from X (a k-depth intelim refutation of X) if $\mathcal{T}$ is an intelim proof of φ from X (an intelim refutation of X) and its depth is k.

Definition 6 *For all X, φ, (i) φ is 0-depth deducible from X, $X \vdash_0 \varphi$, iff there is a 0-depth intelim proof of φ from X; (ii) X is 0-depth refutable, $X \vdash_0$, iff there is a 0-depth intelim refutation of X.*

We shall abuse of the same relation symbol '$\vdash_0$' to denote 0-depth deducibility and refutability. It is easy to show that:

Proposition 2 $\vdash_0$ *is a Tarskian consequence relation.*

Definition 7 *For all X, φ, and for $k > 0$,*

$X \vdash_k \varphi$ iff $X \cup \{\psi\} \vdash_{k-1} \varphi$ and $X \cup \{\bar\psi\} \vdash_{k-1} \varphi$ for some $\psi^u \in$ sub$(X^u \cup \{\varphi^u\})$.

When $X \vdash_k \varphi$, we say that φ is deducible at depth k from X. The above definition covers also the case of k-depth refutability by assuming $X \vdash_k$ as equivalent to $X \vdash_k \varphi$ for all φ. When $X \vdash_k$, we say that X is refutable at depth k.

We shall abuse of the same relation symbol '$\vdash_k$' to denote k-depth deducibility and refutability. Now, it follows immediately from Definitions 5 and 7 that:

Proposition 3 *For all X, φ, $X \vdash_k \varphi$ ($X \vdash_k$) iff there is a k-depth proof of φ from X (a k-depth refutation of X) s.t. all its PB-formulae and PB^*-formulae are in* sub$(X^u \cup \{\varphi^u\})$ (sub(X^u)).

As is the case for **CPL**:

Proposition 4 *The k-depth deducibility relations $\vdash_k$ satisfy reflexivity, monotonicity, but not cut.*

However, it is easy to verify that the relations $\vdash_k$ satisfy the following version of cut:

Depth-bounded cut: If $X \vdash_j \varphi$ and $X \cup \{\varphi\} \vdash_k \psi$, then $X \vdash_{j+k} \psi$.

Further, since $\vdash_0$ is monotonic, its successors are ordered: $\vdash_j \subseteq \vdash_k$ whenever $j \leq k$. The transition from $\vdash_k$ to $\vdash_{k+1}$ corresponds to an increase in the depth at which the nested use of virtual information is allowed. From the adequacy of the unbounded system and the subformula property, it immediately follows that

Proposition 5 $X \vdash \varphi$ in **FDE** iff $X \vdash_k \varphi$ for some k.

We conclude by observing that the decision problem for the k-depth logics is tractable. Theorem 1 immediately suggests a decision procedure for k-depth deducibility: to establish whether φ is k-depth deducible from a finite set X we apply the intelim rules, together with PB and PB^* up to a number k of times, in all possible ways starting from X and restricting to applications which preserve the subformula property. If the resulting intelim tree is closed or φ occurs at the end of all its open branches, then φ is k-depth deducible from X, otherwise it is not. We shall first show the tractability of the 0-depth logic, and then the tractability of the k-depth logics, $k > 0$.

Theorem 2 Whether or not $X \vdash_0 \varphi$ ($X \vdash_0$) can be decided in time $O(n^2)$, where n is the total number of occurrences of symbols in the elements of $X \cup \{\varphi\}$ (of X).

Proof sketch. The proof can be adapted from (D'Agostino et al., 2013). We just sketch the decision procedure and give a hint about the upper bound.

We now describe a general procedure to generate the closure of a set Y of signed formulae under the intelim rules restricting our attention to a finite search space Δ that includes all the formulae in Y^u and is closed under subformulae. Start by constructing the subformula graph associated with Δ, i.e., the graph in which the nodes are the subformulae of Δ, while the edges represent the subformula relation. Observe that the number of distinct subformulae of a formula is always less than or equal to the number of occurrences of symbols in that formula. So the number of distinct subformulae of the formulae in Y is $O(n)$, where n is the number of occurrences of symbols in the elements of Y. Constructing this graph takes time $O(n^2)$. A *neighbour* of a node A is a node consisting of either (i) one of the immediate subformulae of A (if any), or (ii) one of the immediate superformulae of A (if any), or (iii) else one of the immediate subformulae of the immediate superformulae of A (if any). The number of neighbours of each node is $O(n)$.

Let us say that a node in the subformula graph is associated with a premise (a conclusion) of an intelim rule, if it consists of a formula that is the unsigned

Tractable Depth-Bounded Approximations to FDE

part of a premise or of the conclusion. Note that

The relation "A is a neighbour of B" is symmetric. (1)

The node associated with a premise of an intelim rule is a neighbour (2)
both of the node associated with the second premise (if any) and of
the node associated with the conclusion.

Nodes are labelled with a subset of the four signs as follows. Initially, all nodes are marked as "fulfilled". Whenever a new sign is added to the labelling set, the node turns "unfulfilled". At the beginning all the nodes consisting of the formulae in X^u are labelled in accordance with their signs in X (and therefore turn "unfulfilled") while all the others are labelled with the empty set. Fulfilling a node means that all the possible intelim rules involving this node and any of its neighbours are applied, which may lead to adding new signs to the labelling sets of the nodes in the neighbours, making them unfulfilled. This amounts to using the formula in the node to be fulfilled, prefixed with each of the signs in its labelling set, as premise of an intelim rule, possibly involving one of its neighbours as second premise. Yet-unfulfilled nodes are fulfilled in turn (the order is immaterial) and marked as such. Since there are $O(n)$ neighbours, fulfilling a node takes $O(n)$ steps.

A node is *inconsistent* if its label contains a pair of conjugate signs, otherwise it is *consistent*. Note that the labelling set of each consistent node may contain at most two signs. Note also that it may be necessary to fulfil a node more than once, when a *new* sign is added to its labelling set as a result of an application of an intelim rule to one of its neighbours. However, no consistent node needs to be fulfilled more than twice (once for each sign in its labelling set). To see this, observe that if the procedure leads to adding a new sign to a node n' (e.g., the sign F to A) that may, in turn, be used together with a previously fulfilled neighbour n (e.g., $A \vee B$ signed with T) as premise of an intelim rule, then n' turns unfulfilled, and the rule in question will be applied anyway when fulfilling n'. For, n is a neighbour of n', by (1), and so is the node n'' consisting of the conclusion of the rule application (B), by (2), whose labelling set will be updated accordingly (adding T).

A *graph* is *inconsistent* if it contains an inconsistent node, otherwise it is *consistent*. In turn, a graph is *0-depth saturated* if it is either inconsistent or it is consistent and each node is marked as fulfilled. A 0-depth saturated graph is obtained in $O(n^2)$ steps, since there are $O(n)$ nodes in the graph, each node is fulfilled at most twice and fulfilling a node takes $O(n)$ steps. Figure 1 shows the initialized graph for the set $X = \{\mathsf{T}\, C \vee (A \vee B), \mathsf{F}\, C, \mathsf{F}\, A \vee (B \wedge \neg C)\}$.

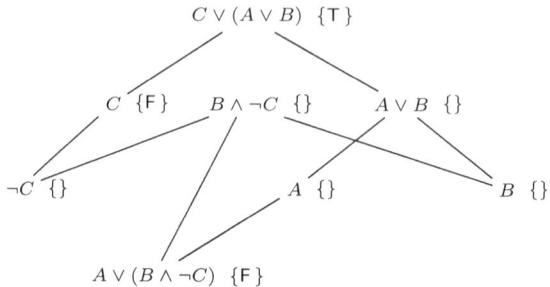

Figure 1: Initialized graph

The corresponding saturated graph, with a possible order of fulfillment of the nodes, is shown in Figure 2. The reader can verify that any alternative sequence leads to the same saturated graph. Figure 3 shows the initialized graph for the set $X = \{\mathsf{T}\, A \vee (B \wedge C), \mathsf{F}\,(A \vee B) \wedge (A \vee C), \mathsf{F}\, A\}$. A corresponding saturated graph, with a possible order of fulfillment of the nodes, is shown in Figure 4. Note that, for inconsistent graphs, not any alternative sequence leads to the same saturated graph. In general, all the signed formulae φ of the form $\mathsf{S}\, A$, where S is in the labelling set of A, that occur in a saturated graph are 0-depth deducible from X.

To decide whether $X \vdash_0 \varphi$ ($X \vdash_0$), consider the graph associated with $X^u \cup \varphi^u$ (X^u), initialize it by adding signs to the labelling sets in accordance with X, and then run the saturation procedure. When the graph is saturated $X \vdash_0 \varphi$ iff the sign of the signed formula φ belongs to the labelling set of φ^u or the graph is inconsistent. Note that an inconsistent graph detects a "metalevel" inconsistency that concerns an incoherent assignment of the imprecise values associated with the signs. Note also that a 0-depth saturated graph starting with nodes labelled with $\{\mathsf{T}\}$ is always consistent and may contain only the signs T and F^* in the labelling sets. □

Corollary 1 *Whether or not $X \vdash_k \varphi$ ($X \vdash_k$) can be decided in time $O(n^{k+2})$, where n is the total number of occurrences of symbols in the elements of $X \cup \{\varphi\}$ (of X).*

Hint. From Definition 7 and the observation that there are $O(n)$ distinct subformulae of $X^u \cup \{\varphi^u\}$ (X^u).

□

Tractable Depth-Bounded Approximations to FDE

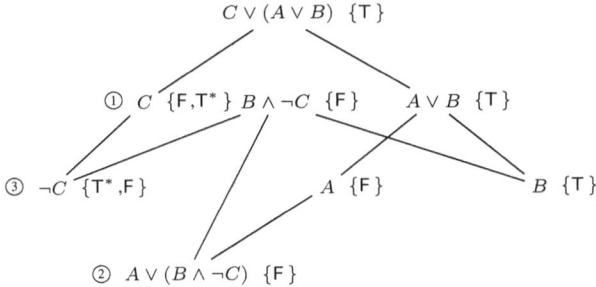

Figure 2: Saturated graph

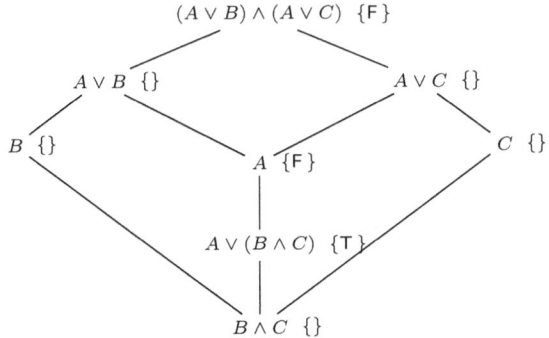

Figure 3: Initialized graph

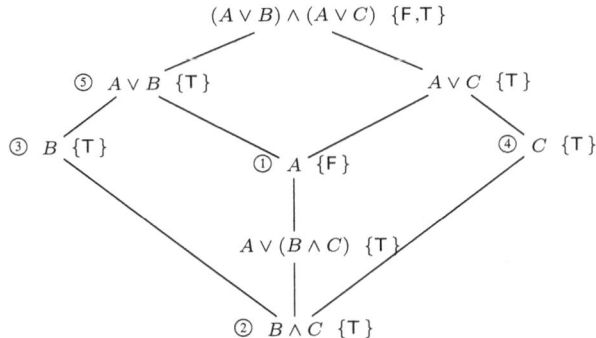

Figure 4: Saturated graph

6 Conclusions

We approached **FDE** via a proof system formulated by means of signed formulae, where the signs express imprecise values associated with two distinct bipartitions of the set of standard 4 values. The proof system consists of linear intelim rules and two branching rules expressing a generalized rule of bivalence, and naturally leads to defining an infinite hierarchy of tractable approximations to **FDE** by controlling the application of the latter rules. Unlike the intelim rules, the branching rules introduce hypothetical information about the imprecise value of a formula. Intuitively, the more virtual information needs to be invoked via the branching rules, the harder the inference is for an agent. So, the nested application of those rules provides a sensible measure of inferential depth.

In a subsequent paper we shall show that our hierarchy of approximations to **FDE** admits of an intuitive 5-valued non-deterministic semantics. This semantics essentially takes the signs as imprecise values (i.e., two-element sets of the standard values), and a fifth value is introduced to represent the case where the agent's information is insufficient to even establish any of the imprecise values. Further, following the methodology used in this paper, we shall define analogous hierarchies of depth-bounded approximations to two logics closely related to **FDE**, namely, Priest's Logic of Paradox and the Strong Kleene Logic.

References

Anderson, A., & Belnap, N. (1962). Tautological entailments. *Philosophical Studies*, *13*(1-2), 9–24.

Arieli, O., Avron, A., & Zamansky, A. (2018). *Theory of Effective Propositional Paraconsistent Logics*. London: College Publications.

Arieli, O., & Denecker, M. (2003). Reducing preferential paraconsistent reasoning to classical entailment. *Journal of Logic and Computation*, *13*(4), 557–580.

Avron, A. (2003). Tableaux with four signs as a unified framework. In M. Cialdea Mayer & F. Pirri (Eds.), *Automated Reasoning with Analytic Tableaux and Related Methods, Int. Conf., TABLEAUX 2003* (Vol. 2796, pp. 4–16). Berlin, Heidelberg: Springer.

Belnap, N. (1977a). How a computer should think. In G. Ryle (Ed.), *Contemporary Aspects of Philosophy* (pp. 30–55). Stocksfield: Oriel Press.

Belnap, N. (1977b). A useful four-valued logic. In J. M. Dunn & G. Epstein (Eds.), *Modern Uses of Multiple-Valued Logics* (pp. 5–37). Dordrecht: Reidel Publishing Company.

Blasio, C. (2015). Up the hill: on the notion of information in logics based on the four-valued bilattice. In *2nd Workshop on Philosophy, Logic and Analytical Metaphysics (FILOMENA 2)* (p. 99).

Blasio, C. (2017). Revisitando a lógica de Dunn-Belnap. *Manuscrito*, *40*(2), 99–126.

Caleiro, C., Marcos, J., & Volpe, M. (2015). Bivalent semantics, generalized compositionality and analytic classic-like tableaux for finite-valued logics. *Theoretical Computer Science*, *603*, 84–110.

Carnielli, W. (1987). Systematization of finite many-valued logics through the method of tableaux. *The Journal of Symbolic Logic*, *52*(2), 473–493.

Cook, S. (1971). The complexity of theorem-proving procedures. In *Proceedings of the Third Annual ACM Symposium on Theory of Computing* (pp. 151–158). New York, NY: Association for Computing Machinery.

D'Agostino, M. (1990). *Investigations into the Complexity of some Propositional Calculi*. Oxford University. Computing Laboratory. Programming Research Group.

D'Agostino, M. (2014). Analytic inference and the informational meaning of the logical operators. *Logique et Analyse*, *57*(227), 407–437.

D'Agostino, M. (2015). An informational view of classical logic. *Theoretical Computer Science*, *606*, 79–97.

D'Agostino, M., Finger, M., & Gabbay, D. (2013). Semantics and proof-theory of depth bounded Boolean logics. *Theoretical Computer Science*, *480*, 43–68.

D'Agostino, M., & Floridi, L. (2009). The enduring scandal of deduction. *Synthese*, *167*(2), 271–315.

D'Agostino, M., Gabbay, D., & Broda, K. (1999). Tableau methods for substructural logics. In M. D'Agostino, D. M. Gabbay, R. Hähnle, & J. Posegga (Eds.), *Handbook of Tableau Methods* (pp. 397–467). Dordrecht: Springer.

D'Agostino, M., Larese, C., & Modgil, S. (2021). Towards depth-bounded natural deduction for classical first-order logic. *Journal of Applied Logics*, *8*(2), 423–451.

Dunn, J. (1976). Intuitive semantics for first-degree entailments and 'coupled trees'. *Philosophical Studies*, *29*(3), 149–168.

Fitting, M. (1991). Bilattices and the semantics of logic programming. *The Journal of Logic Programming*, *11*(2), 91–116.

Fitting, M. (1994). Kleene's three valued logics and their children. *Fundamenta Informaticae*, *20*(1, 2, 3), 113–131.

Hähnle, R. (1999). Tableaux for many-valued logics. In M. D'Agostino, D. M. Gabbay, R. Hähnle, & J. Posegga (Eds.), *Handbook of Tableau Methods* (pp. 529–580). Dordrecht: Springer.

Omori, H., & Wansing, H. (2017). 40 years of FDE: An introductory overview. *Studia Logica*, *105*(6), 1021–1049.

Priest, G. (2001). *An Introduction to Non-Classical Logic*. Cambridge: Cambridge University Press.

Quine, W. (1973). *The Roots of Reference*. LaSalle: Open Court.

Shramko, Y., & Wansing, H. (2005). Some useful 16-valued logics: How a computer network should think. *Journal of Philosophical Logic*, *34*(2), 121–153.

Urquhart, A. (1990). The complexity of decision procedures in relevance logic. In J. M. Dunn & A. Gupta (Eds.), *Truth or Consequences: Essays in Honor of Nuel Belnap* (pp. 61–76). Dordrecht: Springer Netherlands.

Wansing, H., & Belnap, N. (2010). Generalized truth values. A reply to Dubois. *Logic Journal of IGPL*, *18*(6), 921–935.

Marcello D'Agostino
University of Milan, Department of Philosophy
Italy
E-mail: `marcello.dagostino@unimi.it`

Alejandro Solares-Rojas
University of Milan, Department of Philosophy
Italy
E-mail: `alejandro.solares@unimi.it`

Revisiting Brandom's Incompatibility Semantics

CHRISTIAN G. FERMÜLLER AND JOHANNES HAFNER[1]

Abstract: We critically analyse Robert Brandom's incompatibility semantics for classical S5, developed in the context of analytic pragmatism. Among other problems, we point out that Brandom's claim that incompatibility semantics, although holistic and non-compositional, is nevertheless recursively projectible, rests on an assumption that is at odds with intended applications. We also explore an alternative approach that aims at a formal model of Brandom's concept of a 'game of giving and asking for reasons' (GOGAR).

Keywords: semantics, incompatibility, holism, Robert Brandom

1 Introduction

In his John Locke Lectures, published as *Between Saying and Doing: Towards an Analytic Pragmatism* (2008), Brandom stakes out an ambitious program in the philosophy of language highlighting a pragmatist and inferentialist approach to meaning, that acknowledges the precedence of deontic normative over non-modal vocabulary in the elaboration of successful communication. This endeavour entails a new type of formal semantics for propositional modal logic. This *incompatibility semantics* features several aspects that distinguish it from traditional Tarski/Kripke-style semantics. Rather than truth in a model, the basic notion is that of incompatibility between or, more generally, *incoherence* among a set of sentences. This leads to a *holistic* account, in which the semantic status of particular sentences can only be asserted relative to a given context of other sentences. As a consequence the semantics is *non-compositional*: the meaning of a logically complex sentence is not determined by the semantic interpretants (incompatibilities) of just its parts and the connectives used to form it. It is often claimed that holistic, non-compositional semantics cannot account for the projectibility and systematicity of language, and hence also not for

---

[1] Both authors are supported by FWF project P32684-N.

its learnability. Therefore it is at the core of Brandom's project to address such worries by showing that incompatibility semantics enjoys what he calls *'recursive projectibility'*.

The paper is organized as follows. After briefly discussing its main tenets and presenting its formal ingredients in Section 2, we take a critical look at some aspects of Brandom's presentation in Section 3 and suggest some corresponding amendments. In Section 4, we point out that Brandom's claim that incompatibility semantics admits recursive projectibility rests on an additional assumption, namely *'inferential conservativity'*, that is not entailed by his axioms and that is actually quite problematic with respect to the intended models.[2] Section 5 seeks to clarify the relation between incompatibility semantics and classical logic, including its extension to modal logic S5, in a manner that is more transparent than Brandom's own take. Since incompatibility semantics, as presented by Brandom (2008), is largely severed from its pragmatist and inferentialist context, we suggest in Section 6 an alternative approach by introducing a formal model of the *game of giving and asking for reason* (GOGAR), introduced by Brandom (1994), that interprets the inference rules of a particular sequent calculus for intuitionistic logic as interactions between a proponent (of a claim) and a questioner. We conclude in Section 7 with some remarks about open issues and the relation to more recent work by Brandom and his collaborators.

2 Brandom's incompatibility semantics in a nutshell

In the following we are working with languages $\mathcal{L}$ for propositional modal logic. Each language is a set of atomic or logically complex sentences.[3] Such languages need not be syntactically closed "upwards", i.e. in terms of forming arbitrarily complex sentences by applying logical connectives. However, all languages are assumed to be closed "downwards", such that all subformulas of sentences in $\mathcal{L}$ are themselves members of $\mathcal{L}$. These languages are called *proper*.

The point of departure for Brandom's semantic theory is his "suggestion: represent the propositional content expressed by a sentence with the set of sentences that express propositions incompatible with it" (Brandom, 2008, p. 123). *Incompatibility* is a material, binary relation among sentences,

---

[2]The problem described in Section 4 has already been outlined, from a somewhat different perspective, in (Fermüller, 2010).

[3]We will use $p, q, r, \ldots$ for atomic formulas and $F, G, H, \ldots$ for arbitrary formulas.

a generalization of contradictoriness to the case of non-logical properties. However, as Brandom is quick to point out, such a semantic representation of propositional content would be too narrow, since it fails to take into account incompatibilities that hold between three or more sentences while any two of them are compatible. To overcome this limitation, Brandom generalizes incompatibility from a relation among sentences to a relation among sets of sentences. Formally, an *incompatibility function* I from the power set of $\mathcal{L}$, $\mathcal{P}(\mathcal{L})$, to $\mathcal{P}(\mathcal{P}(\mathcal{L}))$ relates to each set of sentences the set of sets of sentences that are incompatible with it. An ordered pair $\langle \mathcal{L}, I \rangle$ is called an *standard incompatibility frame* on $\mathcal{L}$. In addition, Brandom introduces the (non-relational) notion of *incoherence*, which is a material generalization of inconsistency applying to (single) sets of sentences. Formally, $Inc \subseteq \mathcal{P}(\mathcal{L})$ comprises all and only the incoherent sets of sentences of a given language $\mathcal{L}$. Incoherence is a property of a set which is inherited by any superset of it, i.e. *Inc* satisfies the following axiom.

Axiom (Persistence) $\forall$ finite $X, Y \subseteq \mathcal{L}$, and $X \subseteq Y$, if $X \in Inc$ then $Y \in Inc$.

An ordered pair $\langle \mathcal{L}, Inc \rangle$ is called a *standard incoherence frame* on $\mathcal{L}$. For any given language $\mathcal{L}$, the standard incompatibility frame and the standard incoherence frame correspond to each other in virtue of jointly satisfying the following axiom.

Axiom (Partition) $\forall X, Y \subseteq \mathcal{L}, X \cup Y \in Inc$ iff $X \in I(Y)$.

Since reference and truth are not among the resources of incompatibility semantics, (material) entailment is not to be defined in terms of truth-preservation. Rather, material[4] entailment can be viewed in terms of the preservation of compatibility or coherence from premises to conclusion, i.e. the entailment relation between premise set and conclusion holds iff every set of sentences that is compatible with the set of premises is also compatible with the conclusion. This idea is equivalently expressed by the notion of *(material) incompatibility-entailment*. A set X of sentences materially incompatibility-entails a sentence F iff every set Z that is incompatible with $\{F\}$, is also incompatible with X. Brandom generalizes this to a set Y in place of a single sentence F in the following way.

---

[4]Brandom is trying to articulate "the material (that is, non-, or better, pre-logical) sense of 'good inference' [in which sense, for instance] 'Pedro is a donkey,' incompatibility-entails 'Pedro is a mammal'" (Brandom, 2008, p. 121).

Christian G. Fermüller and Johannes Hafner

Definition 1 (Incompatibility-Entailment) *Given an incompatibility function I, a (possibly infinite) set $X \subseteq \mathcal{L}$ and a finite set $Y \subseteq \mathcal{L}$,*

$$X \models_I Y \text{ iff } \bigcap_{F \in Y} I(\{F\}) \subseteq I(X).$$

According to this definition, the set Y is read disjunctively, i.e. the "heuristic meaning" of $X \models_I \{F_1, \ldots, F_n\}$ is that X entails F_1 or ... or F_n (Brandom, 2008, p. 42).

Definition 2 (Validity) $\forall X \subseteq \mathcal{L}$, X *is valid iff* $Y \in \bigcap_{F \in X} I(\{F\}) \Rightarrow Y \in Inc$.

X is valid iff only incoherent sets are incompatible with X, read disjunctively. As a special case we have $\emptyset \models_I \{F\}$ iff $\{F\}$ is valid.

Now we are in a position to introduce, axiomatically, the logical operators, negation, conjunction and necessity.

Axiom (Negation Introduction; NI) $\forall X \subseteq \mathcal{L}, X \cup \{\neg F\} \in Inc$ iff $X \models F$.

Axiom (Conjunction Introduction; CI) $\forall X \subseteq \mathcal{L}, X \cup \{F \wedge G\} \in Inc$ iff $X \cup \{F, G\}$.

Axiom ($\Box$ Introduction; LI) $\forall X \subseteq \mathcal{L}, X \cup \{\Box F\} \in Inc$ iff $X \in Inc$ or $\exists Y \subseteq \mathcal{L}[X \cup Y \notin Inc \ \& \ Y \not\models \{F\}]$.

These axioms specify the content expressed by a logically compound sentence, according to incompatibility semantics, by specifying which sets of sentences are incompatible with it. However, it is a crucial feature of this semantics that the semantic interpretants of compound sentences are not determined by or computable from the semantic interpretants of their components alone, i.e. that this semantics is *not compositional*. For instance, it is easy to see that what is incompatible with F, on the one hand, and what is incompatible with G, on the other, do not fully determine what is incompatible with $\{F, G\}$. This non-compositionality makes incompatibility semantics *holistic*. It has been a common criticism of semantic holism that non-compositionality prevents it from accounting for the projectibility, systematicity and learnability of language. Brandom insists that the standard arguments to this effect are fallacious, and are refuted by his incompatibility semantics.

> ...although that semantics is not compositional, it is fully recursive. The semantic values of logically compound expressions are wholly determined by the semantic values of logically simpler ones. It is holistic, that is, non-compositional, [...] But this holism within each level of constructional complexity is entirely compatible with recursiveness between levels. And this is not just a philosophical claim of mine. The system I am describing allows us to prove it. (In this context, proof is the word made flesh.) (Brandom, 2008, p. 135)

In the rest of the paper we are going to critically evaluate Brandom's claim (and his 'proof' of it) as well as essential parts of the underlying machinery of incompatibility semantics.

3 A critical analysis of Brandom's account

To gain a better understanding of the essential features of the suggested semantic framework, we take a closer, critical look at Brandom's definitions and axioms outlined in Section 2. First, note that Brandom employs an unconventional concept of 'language'. While a language $\mathcal{L}$ may contain logically complex formulas, $\mathcal{L}$, in general, is not closed under forming complex formulas from given ones using the logical connectives. We may, however, assume that all languages considered here are *proper*, i.e., they are closed under taking sub-formulas. Moreover, for our purposes, it is sufficient to consider only *finite languages*. A more problematic feature of Brandom's notion of a logical language is that he only considers conjunction ($\wedge$), negation ($\neg$), and the modal operator ($\Box$) for necessity. Disjunction and implication are only introduced as operators 'abbreviating' certain formulas, built up from atomic formulas using conjunction and negation only. While one can, of course, treat all connectives of *classical logic* as defined from conjunction and negation alone, this is not the case for most other logics. If the aim of incompatibility semantics is to provide an independent approach to the meaning of logical connectives that does not exclude nonclassical logics from the outset, then it is certainly odd to declare that $F \vee G$ is to be understood as abbreviation for $\neg(\neg F \wedge \neg G)$ and that also $F \to G$ has no meaning beyond serving as abbreviation for $\neg(F \wedge \neg G)$. In particular, the decision to treat disjunction and implication not as first class citizens of the logical vocabulary, but as defined connectives, excludes the possibility to come up with a notion of logical validity that is able to distinguish between classical and intuitionistic logic, as demonstrated by the following well-known fact.

Fact 1 *Over the language fragment, where $\neg$ and $\wedge$ are the only logical connectives, classical tautologies coincide with intuitionistic tautologies.*

We will indicate in Section 5 how one can augment Brandom's account to accommodate a richer logical vocabulary. More importantly, we will introduce an alternative to incompatibility semantics in Section 6 that will treat the usual propositional connectives on a par with each other and that arguably remains closer to Brandom's own pragmatist and inferentialist perspective.

As pointed out in Section 2, Brandom presents incompatibility semantics in terms of 5 axioms and some accompanying definitions. They center around the notions of *incompatibility* and *incoherence*, the latter modeled as a set *Inc* of sets of formulas, i.e., $\mathit{Inc} \subseteq \mathcal{P}(\mathcal{L})$. The *Axiom of Persistence* postulates *incoherence* to be monotonic (with respect to the subset relation), which induces a monotonic notion of material inference. While there is nothing wrong with investigating a monotonic notion of (material) incoherence, it is somewhat strange that the *Persistence Axiom* explicitly restricts monotonicity to *finite* sets of sentences. Since Brandom hardly wants to claim that infinite sets of sentences may be coherent even if they contain incoherent subsets, we interpret this oddity simply as an indication that the presented version of incompatibility semantics is intended for finite language scenarios only.[5]

Brandom defines a *standard incoherence frame* as a pair $\langle \mathcal{L}, \mathit{Inc} \rangle$. Since this definition refers to the *material* level of an interpreted language and not to a logical frame in the sense of Kripke semantics for modal logic, it would be better to speak of an incoherence *model*, instead. Even more confusingly, there is a second definition of a *standard incompatibility frame* as a pair $\langle \mathcal{L}, \mathrm{I} \rangle$, where I is a function of type $\mathcal{P}(\mathcal{L}) \mapsto \mathcal{P}(\mathcal{P}(\mathcal{L}))$, called *incompatibility function*. Since $X \in \mathrm{I}(Y)$ iff $X \cup Y \in \mathit{Inc}$ (*Partition Axiom*), incoherence frames and incompatibility frames amount to just two different presentations of the same concept. The *Partition Axiom* can be understood as solely introducing convenient notation.

To amend the outlined infelicities in Brandom's presentation of incompatibility semantics, we will adopt the following alternative notion.

---

[5]There remains a minor lacuna in Brandom's account: the empty set should be declared to be coherent ($\emptyset \notin \mathit{Inc}$). While this follows from persistence if there are other coherent sets, one needs to make it explicit for models where all other sets are incoherent.

Definition 3 *An* incoherence model *$Inc_\mathcal{L}$ over the (finite) language $\mathcal{L}$, is a subset of $\mathcal{P}(\mathcal{L})$, such that*

(i) $\emptyset \notin Inc_\mathcal{L}$,

(ii) *if $X \subseteq Y$ and $X \in Inc_\mathcal{L}$, then $Y \in Inc_\mathcal{L}$.*

When there is no danger of confusion, we will drop the subscript referring to $\mathcal{L}$. The sets in *Inc* are called *incoherent* with respect to *Inc*. If $X \notin Inc$ then X is *coherent* with respect to *Inc*. We say that X is *incompatible* with Y (with respect to *Inc*) if $X \cup Y \in Inc$. The set of sets that are incompatible with Y is denoted by $I(Y)$. If $X = \{F\}$ we also say that the formula F is incompatible with Y (in *Inc*).

The remaining three axioms, called **CI**, **NI**, and **LI** in Brandom (2008), concern conjunction, negation, and necessity, respectively (see Section 2). Inspecting Brandom's formulations of these axioms, restated in Section 2, more closely, reveals that in each case one has to restrict the corresponding statements to refer to only those languages in which the exhibited complex formula actually occurs. (Recall that Brandom allows for languages that contain, e.g., F and G, but not $F \wedge G$, $\neg F$ or $\Box F$. For such languages, **CI**, **NI**, and **LI** are inappropriate without a restricting clause.) From now on, we will assume that in any statement that implicitly or explicitly refers to some language $\mathcal{L}$, $\mathcal{L}$ contains every formula that is mentioned in that statement.

The axiom for conjunction (**CI**) just states that, in judging incoherence, sets of formulas are treated as conjunctions of formulas. Since Brandom wants incoherence to serve as 'a generalization of inconsistency to the case of non-logical properties' (Brandom, 2008, p. 141) this is an obvious choice. The axioms for negation and necessity are much less straightforward, since they involve the notation of incompatibility-entailment, defined by $X \models_I Y$ iff $\bigcap_{F \in Y} I(\{F\}) \subseteq I(X)$.[6] Brandom wants Y to be read disjunctively, rather than conjunctively, in order to be able to mimic Gentzen's classical sequent calculus **LK**. While understandable as a proof strategy this does not sit well with the general setup of incompatibility semantics as Brandom himself seems to admit in a footnote on page 123 of (Brandom, 2008), where he explicitly states that a 'very natural way' to generalize from a single-conclusion to a multiple-conclusion version of incompatibility entailment is

---

[6]Right after the definition Brandom adds: 'When Y is empty we read $\bigcap_{F \in Y} I(\{F\})$ as equivalent to $\mathcal{P}(\mathcal{L})$'. However, e.g., the proof of Claim 2.1 (Weakening) (Brandom, 2008, p. 143) starts as follows: 'Suppose $X \models Y$. Then $\bigcap_{F \in Y} I(\{F\}) \subseteq I(X)$.' But this is clearly false if $\bigcap_{F \in Y} I(\{F\})$ is to be read as $\mathcal{P}(\mathcal{L})$, as stipulated in the cited remark. Moreover, identifying $I(\emptyset)$ with $\mathcal{P}(\mathcal{L})$ is already ruled out by the Partition Axiom.

to interpret the set Y on the right hand side as a conjunction. We therefore suggest to consider the following definition, that has the additional advantage of making transparent that one does not need to involve the function I:

Definition 4 *The set of formulas X materially incompatibility-entails the set of formulas Y with respect to an incoherence model $Inc_\mathcal{L}$, written $X \models_{Inc} Y$, iff for all $Z \subseteq \mathcal{L}$: $Y \cup Z \in Inc$ implies $X \cup Z \in Inc$.*

We write $F_1, \ldots, F_n \models G$, instead of $\{F_1, \ldots, F_n\} \models \{G\}$. Note that in this single-conclusion case Definition 4 coincides with Brandom's original definition. We observe that $X \models_{Inc} \emptyset$ holds for every X. This is also the case for Brandom's original definition, if one ignores his additional remark about identifying the empty intersection of sets of formulas with $\mathcal{P}(\mathcal{L})$. If one wants to express incoherence in terms of entailment, then one should introduce the logical constant $\bot$ (*falsum*). Moreover, it is useful to also add $\top$ (*verum*). Brandom's axioms for introducing connectives should consequently be augmented as follows.

(*Falsum* **Introduction** $\bot$**I:**) $\{\bot\} \in Inc$

(*Verum* **Introduction** $\top$**I:**) $X \cup \{\top\} \in Inc$ iff $X \in Inc$.

Note that $\bot$**I** guarantees that $X \in Inc$ is equivalent to $X \models_{Inc} \bot$.

Brandom also defines a notation of 'validity' that renders a formula F 'valid' iff $\emptyset \models_{Inc} \{F\}$. This piece of terminology is unfortunate, since validity traditionally does not refer to the material level of a given interpretation, but rather singles out what holds with respect to *all* interpretations. If one wants to preserve validity as a *logical* notion, one should call F valid iff $\emptyset \models_{Inc} \{F\}$ *for every incoherence model $Inc_\mathcal{L}$*. Note that only with respect to this latter, more traditional notion, does it make sense to claim, as Brandom does, that the 'intrinsic' logic of incompatibility is the classical modal logic S5.

Fortunately, neither Brandom's strange notion of validity, nor his problematic generalized entailment relation matter for the presentation of axioms **NI**, and **LI**, since only single-conclusion material entailment is used there. However, both axioms are highly problematic with respect to the claim that incompatibility semantics enjoys recursive projectibility, by which Brandom means that judgments about the incoherence of sets of logically complex sentences can be systematically reduced to judgments that only involve less complex sentences. To get a better view on the problem, we reformulate **NI** without making explicit use of the entailment relation.

(Negation Introduction NI:) $X \cup \{\neg F\} \in Inc_{\mathcal{L}}$ iff for all $Y \subseteq \mathcal{L}$: if $\{F\} \cup Y \in Inc_{\mathcal{L}}$ then $X \cup Y \in Inc_{\mathcal{L}}$.

This condition is *circular*, which is brought out most clearly by considering the instance in which $X = \emptyset$ and $Y = \{\neg F\}$. Since we may safely assume that $\{F, \neg F\} \in Inc_{\mathcal{L}}$, the statement boils down to $\{\neg F\} \in Inc_{\mathcal{L}}$ iff $\{\neg F\} \in Inc_{\mathcal{L}}$ in this case. [7]

The axiom **LI** for introducing the modal operator is plagued by the same problem as **NI**: the statement is circular. The scope of the quantifier in the statement refers to all sets of formulas in the given language and hence includes the set $X \cup \{\Box F\}$, the (in)coherence of which is to be settled.

Let us sum up our analysis of Brandom's presentation of incompatibility semantics, so far. Brandom's axioms refer to three basic notions: *Inc* (incoherence as a property of sets of formulas), the incompatibility function I, and incompatibility-entailment. In fact, each of these notions can be defined in terms of any of the two other notions. In particular, it is sufficient to consider just *Inc*. Since *Inc* refers to the material level, one should replace Brandom's talk of 'incoherence frames' and 'incompatibility frames' by references to *incoherence models*, as defined in Definition 3. This leaves only the axioms **CI**, **NI**, and **LI** to be considered. The fact that non-modal formulas are built up from atomic formulas using conjunction and negation only, spoils prospects to come up with a complete, independent semantic framework for (potentially) nonclassical logics. Most problematic, however, is the fact that the axioms for negation (**NI**) and for necessity (**LI**) are circular. In the next section, we investigate why Brandom nevertheless thinks that incompatibility semantics admits 'recursive projectibility' and characterizes the logic S5.

---

[7] There seems to be a tension (an incoherence?) between the accounts, respectively, in the main text and the appendix of (Brandom, 2008) concerning the semantics of negated sentences. On the one hand, the main text (p. 127) suggests a recursive, step by step, extension of an incoherence frame in tandem with the corresponding incompatibility consequence relation in order to provide the incompatibility sets for more and more complex negated sentences. Such a stepwise construction of the semantics of negated sentences may avoid circularities. However, on the other hand, in the appendix (p. 142) the semantics of all logical connectives is clearly presented axiomatically and there is no indication of a semantic construction by successive extensions starting from an incoherence frame for a language containing only atomic formulas etc. Such a construction would also require more machinery, such as definitions of how to dovetail the respective extensions of the language/semantics by the three connectives.

Christian G. Fermüller and Johannes Hafner

4 Problems with recursive projectibility

To get a better grip on the circularity problem outlined in the last section, we focus on negation and consider the following example.

Example 1 Let $\mathcal{L} = \{p, q\}$ and $Inc_{\mathcal{L}} = \emptyset$. Obviously $p \models_{Inc} p$ and $q \models_{Inc} q$. Since there are no incoherent sets, we also have $p \models_{Inc} q$ and $q \models_{Inc} p$.

Now let us consider $\mathcal{L}' = \{p, q, \neg p\}$. In Brandom's terminology, $\mathcal{L}'$ is a proper extension of $\mathcal{L}$. Since p entails itself, axiom **NI** yields $\{p, \neg p\} \in Inc'_{\mathcal{L}'}$ for any Inc' over $\mathcal{L}'$. We want to keep $\{p, q\}$ coherent, like in $Inc_{\mathcal{L}}$. But what about $\{q, \neg p\}$? Here we run into the circularity pointed out in Section 3: **NI** requires that $\{q, \neg p\} \in Inc'_{\mathcal{L}'}$ iff for all $Y \subseteq \mathcal{L}'$ either $Y \cup \{p\}$ is coherent or $Y \cup \{q\}$ is incoherent. Since $\{p\}$ and $\{p, q\}$ are coherent and $\{p, \neg p\}$ is incoherent this boils down to $\{q, \neg p\} \in Inc'_{\mathcal{L}'}$ iff $\{q, \neg p\} \in Inc'_{\mathcal{L}'}$. In other words, we are free to declare $\{q, \neg p\}$ to be either coherent or to be incoherent in $Inc'_{\mathcal{L}'}$ without violating Brandom's axioms for incompatibility semantics.

As we have seen in Section 2, a central claim of Brandom is that incompatibility semantics, although holistic, nevertheless is 'fully recursive'. More precisely, Brandom claims:

> The semantic values of all the logically compound sentences are computable entirely from the values of less complex sentences. (Brandom, 2008, p. 135)

But Example 1 refutes this claim. The semantic value of $\neg p$ in the model $Inc'_{\mathcal{L}'}$, i.e. its coherence or incoherence jointly with other sentences, is *not* determined by the coherence or incoherence of the sets of sentences in the language $\mathcal{L}$, where $\mathcal{L}' = \mathcal{L} \cup \{\neg p\}$. So why does Brandom think that he can maintain his claim, although (apparent?) counterexamples are readily specified? The answer to this question can (only) be found in Section 5 of Appendix I to Chapter 5 of (Brandom, 2008). There, it turns out that in extending an incoherence model Inc ('frame' in his terminology) from a language $\mathcal{L}$ to a model Inc' over a richer language $\mathcal{L}'$, just like in Example 1, Brandom does not consider it sufficient that Inc' and Inc coincide over sets of sentences in $\mathcal{L}$. He rather imposes another property, namely *inferential conservativity*, defined as follows.

Revisiting Brandom

Definition 5 *Let $\mathcal{L} \subseteq \mathcal{L}'$ and let Inc be an incoherence model over $\mathcal{L}$. Then an incoherence model Inc' over $\mathcal{L}'$ is* inferentially conservative *with respect to Inc iff, for all $X, Y \subseteq \mathcal{L}$, $X \models_{Inc} Y$ iff $X \models_{Inc'} Y$.*[8]

Brandom shows that for every incoherence model *Inc* over $\mathcal{L}$ and any proper extension $\mathcal{L}'$ of $\mathcal{L}$, there exists a unique smallest incoherence model *Inc'* over $\mathcal{L}'$, called *the model for $\mathcal{L}'$ determined by Inc*, that is inferentially conservative with respect to *Inc*.

Let us revisit Example 1 to see whether the insistence on inferential conservativity indeed settles the question whether $\{q, \neg p\}$ should be incoherent in $Inc'_{\mathcal{L}'}$, extending $Inc_{\mathcal{L}}$. Recall that $q \models_{Inc} p$. If we declare $\{q, \neg p\}$ to be coherent in *Inc'* then we obtain $q \not\models_{Inc'} p$, since $\{p, \neg p\} \in Inc'_{\mathcal{L}'}$, but $\{q, \neg p\} \notin Inc'_{\mathcal{L}'}$. Therefore we have to set $\{q, \neg p\} \in Inc'_{\mathcal{L}'}$, if we want $Inc'_{\mathcal{L}'}$ to be inferentially conservative over $Inc_{\mathcal{L}}$.

We emphasize that Brandom's axioms for incoherence and incompatibility entailment do not constrain the models as indicated above. If incompatibility semantics is intended to apply only to models that are inferentially conservative over a given model over an atomic language, then one should add corresponding axioms. But this complaint should not eclipse a more urgent question, namely whether restricting attention to 'determined' models meshes with the intended meaning and use of incompatibility semantics. We argue that this is not the case, by instantiating the propositional variables p and q of Example 1 with concrete sentences as follows:

p: "It is raining in Vienna."

q: "It is raining in New York."

The incoherence model over the language $\mathcal{L} = \{p, q\}$ in Example 1 declared p to be compatible with q. This is certainly a reasonable choice also with respect to the given natural language interpretation of p and q. Once more, we expand $\mathcal{L}$ to $\mathcal{L}'$ by including also $\neg p$ = "It is not raining in Vienna". As we have seen above, insisting on inferential conservativity forces us to declare "It is not raining in Vienna" to be incompatible with "It is raining in New York". This looks very odd, indeed. Incompatibility semantics should allow us to formally model a situation where not only p is compatible with q, but also $\neg p$ is compatible with q (and where the three mentioned statements

---

[8] In (Brandom, 2008), $\models_{Inc}$ and $\models_{Inc'}$ refer to the disjunctive generalization of the single-conclusion incompatibility entailment relation, whereas we prefer the conjunctive version of Definition 4. However, nothing in our criticism below depends on this choice.

are the only ones under consideration). However, we can only accomplish this if we ignore inferential conservativity and consequently dispense with recursive projectibility.

5 Characterizing classical logic and S5

Brandom (2008) shows that the set of formulas that are coherent in all incoherence models coincides with the set of valid formulas in modal logic S5. The proof is quite involved and not without problems, because of the issue with Brandom's disjunctive version of the generalized incompatibility entailment relation that we have pointed out in Section 3. In any case, it is important to recognize that (material) incompatibility entailment behaves quite differently compared to classical material entailment, even in its ordinary, single-conclusion format. To get a better view of the issue, let's first review some basic notions of classical logic.

Definition 6 *A* (CL-)valuation *(or* interpretation*) is a function v that assigns either 1 (for 'true') or 0 (for 'false') to every propositional variable.*[9] *It is extended to propositional formulas, built up using negation and conjunction, as usual:*

$$v(\neg F) = 1 - v(F) \qquad v(F \wedge G) = \min(v(F), v(G))$$

We may add the atomic formula $\bot$ (falsum), stipulating $v(\bot) = 0$.

A set of formulas X materially CL-entails *a formula F with respect to an interpretation v (written $X \models_v F$) iff $v(G) = 0$ for some $G \in X$ or $v(F) = 1$. X logically* CL-entails *a formula F iff $X \models_v F$ for all interpretations v. F is* CL-valid *($\models_{CL} F$) iff $v(F) = 1$ for all valuations.*

We write $G_1, \ldots, G_n \models_{v/CL} F$ instead of $\{G_1, \ldots, G_n\} \models_{v/CL} F$.

Already on the atomic level, a difference between material CL- and incompatibility entailment emerges. Let p and q be propositional variables. For every valuation v, we clearly have

$$p \models_v q \text{ or } q \models_v p.$$

In contrast, there are incoherence frames *Inc*, such that

$$\text{neither } p \models_{Inc} q \text{ nor } q \models_{Inc} p.$$

---

[9]Usually, one assumes an infinite supply of propositional variables. But when the context fixes a finite language $\mathcal{L}$ (in the sense of Brandom) we may safely assume that only those propositional variables that occur in $\mathcal{L}$ are meant.

Revisiting Brandom

To see the latter, consider a language containing also the propositional variables r and s, and let $\{q, r\} \in \mathit{Inc}$, $\{p, r\} \notin \mathit{Inc}$, $\{p, s\} \in \mathit{Inc}$, $\{q, s\} \notin \mathit{Inc}$.

Even if we restrict attention to languages containing only those propositional variables that are explicitly mentioned in the entailment claim, incompatibility entailment behaves non-classically: if $\mathcal{L} = \{p, q, p \wedge q\}$ and $\{p, q\} \in \mathit{Inc}$, but $\{p\} \notin \mathit{Inc}$ and $\{q\} \notin \mathit{Inc}$ we have

$$p \not\models_{\mathit{Inc}} \bot \text{ and } q \not\models_{\mathit{Inc}} \bot, \text{ but } p \wedge q \models_{\mathit{Inc}} \bot.$$

In spite of the indicated differences, one can establish a clear and tight connection between incompatibility semantics and classical semantics in a manner that is simpler and more transparent than Brandom's approach. The idea is to simply declare a set of sentences to be incoherent iff the conjunction of its members is false. More formally, consider the following correspondence between incompatibility models and sets of CL-valuations.

Definition 7 *Let Inc be an incoherence model over a language $\mathcal{L}$. The corresponding set of CL-valuations $\mathcal{V}_{\mathit{Inc}}$ is defined as*

$$\mathcal{V}_{\mathit{Inc}} = \{v \mid \forall X \in \mathit{Inc}\, \exists F \in X \colon v(F) = 0\}.$$

Since we only consider finite sets of sentences, we have $X \in \mathit{Inc}$ iff all valuations in $\mathcal{V}_{\mathit{Inc}}$ evaluate the conjunction of formulas in X as false.

The inverse translation, from sets of valuations to incoherence models needs to be handled with some care, due to Brandom's unusual definition of a (proper) language $\mathcal{L}$.

Definition 8 *Let $\mathcal{L}$ be a proper language (in the sense of Brandom) and let $\mathcal{V}$ be a set of CL-valuations over the set $\mathcal{L}_{\mathit{at}}$ of propostional variables occurring in $\mathcal{L}$. Then the incoherence model $\mathit{Inc}^{\mathcal{V}}_{\mathcal{L}}$ corresponding to $\mathcal{V}$ is given by*

$$\mathit{Inc}^{\mathcal{V}}_{\mathcal{L}} = \{X \subseteq \mathcal{L} \mid \forall v \in \mathcal{V}, \exists F \in X \colon v(F) = 0\}.$$

Let us indicate why $\mathit{Inc}^{\mathcal{V}}_{\mathcal{L}}$, as just defined, indeed constitutes an incoherence model, complying with Brandom's axioms. Persistence, i.e. the fact that $X \in \mathit{Inc}^{\mathcal{V}}_{\mathcal{L}}$ implies $X' \in \mathit{Inc}^{\mathcal{V}}_{\mathcal{L}}$ for every $X \subseteq X' \subseteq \mathcal{L}$, is maintained since $\exists F \in X \colon v(F) = 0$ trivially implies $\exists F \in X' \colon v(F) = 0$ whenever $X \subseteq X'$.

To see that axiom **CI** holds for $\mathit{Inc}^{\mathcal{V}}_{\mathcal{L}}$, it suffices to recall that we are only dealing with finite languages here and thus may identify a set of formulas with the conjunction of its members. (The empty conjunction is identified with $\top$.)

To understand that also the negation axiom **NI** is satisfied, recall the reformulation of **NI** as stated in Section 3. Since there is no reference to the logical form of the involved formula, except for the negation sign preceding F, we may code X, Y, and F by propositional variables x, y, and f, respectively. In accordance with Definition 8, we may thus expresses the claim $X \cup \{\neg F\} \in \mathit{Inc}_\mathcal{L}^\mathcal{V}$ as the claim that $v(x \wedge \neg f) = 0$ for all corresponding valuations v. Proceeding analogously for $\{F\} \cup Y$ and $X \cup Y$ reduces **NI** to the following claim referring to classical logic:

$$v(x \wedge \neg f) = 0 \text{ iff } \forall y[v(f \wedge y) = 0 \text{ implies } v(x \wedge y) = 0].$$

That this statement is true for all CL-valuations v can be verified by eliminating the propositional quantifier and coding the whole claim as the propositional formula

$$\neg(x \wedge \neg f) \leftrightarrow [(\neg(f \wedge \bot) \to \neg(x \wedge \bot)) \wedge [\neg(f \wedge \top) \to \neg(x \wedge \top))$$

which is a classical tautology.

What about the modal operator $\Box$? We argue that the above analysis straightforwardly generalizes to a characterization of S5. Recall that the standard (Kripke) semantics for modal logics refers to a model $\langle W, R, V \rangle$, where W is a non-empty set of *worlds*, R a binary *accessibility relation* over W, and where V associates a CL-valuation with every $w \in W$. The semantics of $\neg$ and $\wedge$ refers to the CL-valuations as usual, thus assigning a truth value $v_w(F)$ to every non-modal formula F in each world w. This extends to modal formulas via the condition $v_w(\Box F) = 1$ iff $v_{w'}(F) = 1$ for all w' such that $R(w, w')$. A formula F is called *valid* in S5 if for all models, where R is reflexive, symmetric and transitive (i.e. an equivalence relation) $v_w(F) = 1$ in every world w. It is easy to see that only *connected* components of the graph (*frame*) $\langle W, R \rangle$ are relevant when evaluating formulas in a given world. Hence, in the case of S5, we may focus on the special case where R is the *universal relation*, i.e. $R(w, w')$ for all $w, w' \in W$. But in such models duplicates of worlds associated with the same valuation are redundant. To sum up these observations: we may define S5-validity with respect to an arbitrary set of CL-valuations $\mathcal{V}$, rather than with respect to Kripke models, by declaring for any $v \in \mathcal{V}$ that $v(\Box F) = 1$ iff $v(F) = 1$ for all $v \in \mathcal{V}$. Notice that this means that the correspondence between incoherence models and sets of CL-valuations, established by Definitions 7 and 8, carries over to languages that include modal formulas. In other words, we may still interpret $X \in \mathit{Inc}$ as expressing that for every $v \in \mathcal{V}_{\mathit{Inc}}$ we have $v(F) = 0$ for at least one $F \in X$. An argument analogous to that for the axiom **NI**, above, shows that also the axiom **LI** for the introduction of the modal operator $\Box$ is satisfied for the suggested interpretation of incoherence.

6 A GOGAR model

Compared to Brandom's rather involved and indirect proof that S5-validity coincides with coherence in all incoherence models, the considerations in Section 5 provide a much more direct route to the understanding of the relation between incompatibility semantics and classical Tarski/Kripke semantics. However, one might complain that our analysis is at variance with Brandom's deliberate avoidance of reference to truth values and his intention to interpret 'incoherence' as a pragmatic notion, rather than as a purely semantic concept. However, as our above analysis reveals, the setup of the semantic machinery in (Brandom, 2008) is largely severed from Brandom's philosophical stance about logic: accepting incompatibility semantics does not directly support normative inferentialism. In fact we share Brandom's favoring of an approach to logic that gives preference to normative pragmatics over pure semantics and that consequently respects inferentialist insights. We also embrace the concept of logical expressivism regarding the meaning of connectives. Since incompatibility semantics itself hardly adequately meets corresponding demands, we finally want to explore, at least tentatively, an alternative approach that is more directly connected to Brandom's concept of a 'game of giving and asking for reasons' (GOGAR), introduced in *Making It Explicit* (Brandom, 1994) and that may well be classified as inferentialist, pragmatist, and logically expressivist.

We suggest to model GOGAR[10] as a formal game played by a *proponent* **P** and a *questioner* **Q**, reminiscent of the 'dialogical logic' of Lorenzen (1960). The game is not intended to cover the full range of possible interactions between two rational conversationalists, but rather restricts attention to a scenario where **P** seeks to defend a single claim questioned by **Q**. Let us first ignore the logical structure of sentences. The corresponding *atomic game* instantiates the following simple schema:

---

[10] Another formal model of GOGAR is presented in (Porello, 2012).

Christian G. Fermüller and Johannes Hafner

P: *asserts a claim p*
Q: *asks for reasons to accept p*
P: *offers corresponding reasons $r_1, \ldots, r_n$*

At this point **Q** may declare to be satisfied or not. If **Q** is not (yet) satisfied, the game continues by treating the asserted reasons as further claims made by **P** that may be questioned by **Q**. Deciding rationally whether **Q** should declare to be satisfied consists in making two different kinds of judgments: (1) judging whether the reasons $r_1, \ldots, r_n$ materially entail p and (2) judging whether to accept $r_1, \ldots, r_n$ (independently of the claim p). Note that we did not exclude the possibility that **P** offers the asserted claim itself among the reasons to accept it. Such a move by **P** clearly settles the entailment judgment (1), but has no bearing on the second judgment (2). Here, we are only interested in the 'intrinsic logic' of GOGAR and hence focus on the entailment question. At a first glance, this seems to trivialize to task of **P**: simply repeating the claim when questioned by **Q** settles the matter. But this is only the case if we understand the (atomic) game as strictly adversarial, i.e., formally, as a win-lose game. However, we may just as well assume that it is in the interest of the players to make non-trivial entailment claims explicit. In fact, rather than simply referring to an arbitrarily given atomic relation, we may connect the game with incompatibility semantics and stipulate that **P** and **Q** agree about an *atomic* incoherence frame $Inc_{\mathcal{L}}$, where $\mathcal{L}$ is the set of (atomic) sentences assertible as claims or reasons. In this scenario the above game may proceed as follows.

Q: *chooses $q_1, \ldots, q_m$ such that $\{p, q_1, \ldots, q_m\} \in Inc_{\mathcal{L}}$*
P: *replies by pointing out that $\{r_1, \ldots, r_n, q_1, \ldots, q_m\} \in Inc_{\mathcal{L}}$*

Depending on $Inc_{\mathcal{L}}$, **Q** may not be able to make the indicated move. In this case $r_1, \ldots, r_n \models_{Inc} p$ has been established. Similarly, **P** may not be able to make her move, which means that $r_1, \ldots, r_n \not\models_{Inc} p$. Furthermore, if **P** has a reply to every possible move of **Q**, then, again, $r_1, \ldots, r_n \models_{Inc} p$.

Let us now consider richer languages. Like in the atomic case, a general GOGAR instance starts with a claim F by **P**, followed by reasons $G_1, \ldots, G_n$ for accepting F, asserted by **P** after questioned by **Q**. This results in a state denoted by $G_1, \ldots, G_n \triangleright F$, corresponding to a material entailment claim $G_1, \ldots, G_n \models F$. Not only the claim (consequent) F, but also the stated reasons (premises) may be questioned by **Q**. Since we do not want to prevent **Q** from questioning the same sentence more than once during the run of the game, the collection $G_1, \ldots, G_n$ is formally modeled as a *multiset*, rather than a set of sentences.

Revisiting Brandom

In contrast to Brandom, we aim at an autonomous semantics for implication ($\rightarrow$) that does not define $F \rightarrow G$ as an abbreviation of $\neg(F \wedge \neg G)$. Rather, we want to come up with a GOGAR rule that matches the logically expressivist insight that $F \rightarrow G$ is just a syntactic device that allows one to express that G is materially entailed by F in a given context of further assertions. We therefore suggest the following rule that refers to state $H_1, \ldots, H_m \triangleright F \rightarrow G$.

Q: *asks for reasons to accept $F \rightarrow G$ if $H_1, \ldots, H_m$ are accepted*
P: *asserts that $H_1, \ldots, H_m$, augmented by F, are reasons for asserting G*

From now on we will denote such a rule more concisely as

$$\frac{H_1, \ldots, H_m \triangleright F \rightarrow G}{H_1, \ldots, H_m, F \triangleright G}$$

For conjunctive claims we stipulate the following rule.

$$\frac{H_1, \ldots, H_m \triangleright F \wedge G}{H_1, \ldots, H_m \triangleright F \text{ and } H_1, \ldots, H_m \triangleright G}$$

In words: when **Q** asks **P** for reasons to accept $F \wedge G$ given the sentences $H_1, \ldots, H_m$, **P** replies that $H_1, \ldots, H_m$ constitute reasons to accept F as well as reasons to accept G. Thus conjunction ($\wedge$) is treated as an expressive device to join two separate entailment claims. Note that this rule indicates that an overall state of GOGAR is given by a multiset $\{\Gamma_1 \triangleright F_1, \ldots, \Gamma_n \triangleright F_n\}$ of component states, where each Γ_i is a multiset of sentences currently offered as reasons for the claim F_i. To determine how the game is to proceed, we let **Q** choose the component state (entailment claim) to which the next round of interactions has to apply.

The GOGAR rule for disjunction involves a choice by player **P**:

$$\frac{H_1, \ldots, H_m \triangleright F \vee G}{H_1, \ldots, H_m \triangleright F \text{ or } H_1, \ldots, H_m \triangleright G}$$

In other words, in reply to **Q**'s questioning of the claim $F \vee G$, **P** reduces her disjunctive claim to claiming one of the disjuncts.

We still need rules for reducing states that involve logically complex sentences asserted as reasons. The case for conjunctive reasons is particularly simple: if **Q** questions a conjunctive sentence $F \wedge G$, asserted by **P** as a reason for accepting some claim, then **P** simply replaces $F \wedge G$ by F and G in the multiset of asserted reasons. In our concise notation this amounts to

$$\frac{F \wedge G, H_1, \ldots, H_m \triangleright I}{F, G, H_1, \ldots, H_m \triangleright I}$$

For disjunctive reasons we introduce the following rule:

$$\frac{F \vee G, H_1, \ldots, H_m \triangleright I}{F, H_1, \ldots, H_m \triangleright I \text{ and } G, H_1, \ldots, H_m \triangleright I}$$

In other words, if prompted by **Q**, **P** makes explicit that in order to establish the material entailment claim $F \vee G, H_1, \ldots, H_m \models I$ it is sufficient to establish the two entailment claims that result from replacing $F \vee G$ by either F or G, respectively, in the premises (i.e., in the multiset of reasons given so far for accepting I).

Finally, consider the following rule for the case, where **Q** questions $F \to G$ among the reasons given by **P** for accepting the sentence I.

$$\frac{F \to G, H_1, \ldots, H_m \triangleright I}{F \to G, H_1, \ldots, H_m \triangleright F \text{ and } G, H_1, \ldots, H_m \triangleright I}$$

Again, the rule can be understood as making explicit that establishing $F \to G, H_1, \ldots, H_m \models I$ can be reduced to establishing $G, H_1, \ldots, H_m \models I$ as well as $F \to G, H_1, \ldots, H_m \models F$. The presence of $F \to G$ in the latter entailment claim may appear redundant at first glance. However, it turns out that one should allow **Q** to question $F \to G$ again in this situation. (Alternatively, one may introduce a rule forcing **P** to provide another copy of the indicated implication, if prompted to do so by **Q**.)

No separate rules for negated sentences as reasons or claims are needed if $\neg F$ is treated as $F \to \bot$, where $\bot$ denotes a sentence that is incoherent in all interpretations.

To turn our GOGAR model into an ordinary two-person extensive form game, we still have to do three things: (1) define how the game ends, (2) specify pay-off values for both players at final states, and (3) settle a remaining indeterminacy about the possible continuation at non-final states: once **Q** has picked a component (entailment claim) to which one of the above game rules is to be applied, who gets to choose which non-atomic formula (claim or reason) is to be reduced?

Let us address task (1) first. Note that it is not reasonable to declare that the game ends if and only if all sentences in all entailment claims that constitute the current state game are atomic, since we keep non-atomic sentences asserted as reasons available for further questioning. Rather, we declare that the game ends as soon as, in every component (entailment

Revisiting Brandom

claim) of the current state, the claimed sentence (conclusion) is atomic and also appears among the sentences asserted as reasons for it or else $\bot$ appears among the asserted reasons (premises). I.e., final states are multisets of component states, each of the form $p, H_1, \ldots, H_n \triangleright p$ or of the form $\bot, H_1, \ldots, H_n \triangleright p$, where p is an atomic sentence. Of course, there is still no guarantee that a given instance of the game ever ends: we have to take into account infinite runs as well.

Tasks (2) and (3) are not independent of each other. If we model GOGAR as a win-lose game, then we stipulate that **P** wins the game whenever it reaches a final state and **Q** wins if the game runs forever. In this case we should give **P** the right to choose the next non-atomic formula to be reduced (i.e., the rule to be applied) in the component of the current state that has been chosen by **Q**, since otherwise **Q** could trivially force each non-atomic instance of the game to run forever. An alternative option is to model GOGAR as cooperative game, where both players prefer to reach a final state, rather then to play forever. In the latter case, it does not matter whether **P** or **Q** chooses the sentence to be reduced next.

Claim 1 *Suppose that GOGAR starts with **P**'s claim that F and **P** offers the reasons $H_1, \ldots, H_n$ for accepting F, when questioned by **Q**.*

*In the win-lose version of GOGAR, **P** has a winning strategy iff F is a logical consequence of $H_1, \ldots, H_n$ according to intuitionistic logic ($H_1, \ldots, H_n \models_{IL} F$).*

*In the cooperative version of GOGAR, **P** and **Q**, jointly have a winning strategy iff $H_1, \ldots, H_n \models_{IL} F$.*

A detailed proof of Claim 1 is beyond the scope of this paper. However, we may indicate the essence of the proof by directing the reader to the sequent calculus **G3i** for intuitionistic logic in (Troelstra & Schwichtenberg, 2000). Reading our game rules from bottom to top and replacing $\triangleright$ by the sequent arrow turns them into the propositional rules of **G3i**. Moreover, the axioms of **G3i** match the definition of final game states. In this manner, our version of GOGAR emerges as an interpretation of this intuitionistic calculus.[11]

---

[11] It is easy to see that **G3i** proofs correspond to winning strategies in GOGAR. To complete the proof, one still has to establish that the above stipulations about the possible successions of moves amount, in both versions of the game, to a successful proof search strategy in **G3i**.

7 Conclusion

Our re-assessment of incompatibility semantics revealed a number of problems with Brandom's definitions and claims. Most importantly, the central claim that, although holistic and non-compositional, incompatibility semantics admits recursive projectibility and hence refutes the claim that holistic semantics cannot account for the systematicity and learnability of language, has been shown to rest on the additional assumption of inferential conservativity. This assumption, however, is at odds with the intended application of incoherence models as pointed out in Section 4: recursive projectibility can only be obtained for a price that, arguably, is too high to pay.

Another complaint about Brandom's approach to logical semantics is the fact that it does not, at least not directly, amount to a pragmatist and inferentialist account that ties in with logical expressivism. In particular, Brandom's own insight that the logical connective in $F \to G$ can be seen as an expressive device for expressing that G is materially entailed by F is not reflected in a corresponding meaning postulate in incompatibility semantics. Therefore we suggested an alternative approach that seeks to define the meaning of logical connectives via rules of an idealized, formal 'game of giving and asking for reasons', in which a proponent **P** systematically reduces a logically complex entailment claim to an atomic entailment, prompted by systematic questioning by a second player **Q**. While this is similar to dialogical logic (Lorenzen, 1960), a main difference is that the role of **Q** need not necessarily be understood as antagonistic to that of **P**.[12] Moreover, we indicated how our GOGAR model can be connected to the concept of incompatibility-entailment at the atomic level. The game amounts to an interpretation of a cut-free sequent calculus that is sound and complete for intuitionistic logic. It remains to be investigated whether this model can be extended to rules for asserting sentences that feature a modal operator. Another line for further research is to explore connections to the notion of scorekeeping in language games by Lewis (1979) and observations about the necessity of postulating a common ground between effective conversationalists (Stalnaker, 2002). Finally, it should be mentioned that Brandom and his collaborators recently shifted attention to an account of material and logical entailment that, in contrast to the incompatibility semantics of Brandom (2008), embraces *non-monotonicity* and consequently relates to sequent systems that do not admit weakening (see Brandom, 2021). We plan

---

[12] Another major difference is that in the GOGAR model **Q**, unlike the opponent to **P** in Lorenzen's game, does not assert sentences herself, but only questions those asserted by **P**.

to investigate whether our game semantic approach can be extended to cover also corresponding insights about non-monotonic inference.

References

Brandom, R. (1994). *Making It Explicit: Reasoning, Representing, and Discursive Commitment*. Cambridge, MA: Harvard University Press.

Brandom, R. (2008). *Between Saying and Doing: Towards an Analytic Pragmatism*. Oxford: Oxford University Press.

Brandom, R. (2021). *On the structure of reasons: pragmatics, semantics, and logic*. Accessible at https://sites.pitt.edu/~rbrandom.

Fermüller, C. G. (2010). Some critical remarks on incompatibilty semantics. In M. Peliš (Ed.), *Logica Yearbook 2009* (pp. 81–95). London: College Publications.

Lewis, D. (1979). Scorekeeping in a language game. In R. Bäuerle, U. Egli, & A. Stechow (Eds.), *Semantics From Different Points of View* (pp. 172–187). Berlin, Heidelberg: Springer.

Lorenzen, P. (1960). Logik und Agon. In *Atti del XII Congresso Internazionale di Filosofia* (pp. 187–194). Florence: Sansoni.

Porello, D. (2012). Incompatibility semantics from agreement. *Philosophia*, *40*(1), 99–119.

Stalnaker, R. (2002). Common ground. *Linguistics and Philosophy*, *25*(5/6), 701–721.

Troelstra, A. S., & Schwichtenberg, H. (2000). *Basic Proof Theory*. Cambridge: Cambridge University Press.

Christian G. Fermüller
TU Wien, Institute of Logic and Computation
Austria
E-mail: chrisf@logic.at

Johannes Hafner
TU Wien, Institute of Logic and Computation
Austria
E-mail: jhafner.vienna@gmail.com

Reasoning in Commutative Kleene Algebras from *-Free Hypotheses

STEPAN L. KUZNETSOV[1]

Abstract: We prove that reasoning from *-free hypotheses in commutative *-continuous Kleene algebras in Π_2^0-complete, thus providing a commutative counterpart to a result by Kozen (2002).

Keywords: Kleene algebra, commutative Kleene algebra, complexity, reasoning from hypotheses

1 Introduction

Iteration, or Kleene star (Kleene, 1956), is one of the most interesting algebraic operations in computer science. The most basic algebraic structures which employ Kleene star are *Kleene algebras,* defined as follows:

Definition 1 *A Kleene algebra is a structure* $\mathcal{A} = (A; \preceq, +, \cdot, 0, 1, {}^*)$, *where:*

1. $(A; +, \cdot, 0, 1)$ *is an idempotent semiring;*

2. *the partial order $\preceq$ is defined using $+$ as a semilattice structure: $a \preceq b$ iff $a + b = b$;*

3. *iteration obeys two least-fixpoint conditions simultaneously:*

$$a^* = \min_{\preceq}\{b \mid 1 + a \cdot b \preceq b\} = \min_{\preceq}\{b \mid 1 + b \cdot a \preceq b\}.$$

We also define two important subclasses of Kleene algebras.

Definition 2 *Kleene algebra $\mathcal{A}$ is *-continuous, if for any $a, b, c \in A$*

$$b \cdot a^* \cdot c = \sup_{\preceq}\{b \cdot a^n \cdot c \mid n \geq 0\}$$

(here $a^n = \underbrace{a \cdot \ldots \cdot a}_{n \text{ times}}$ for $n > 0$ and $a^0 = 1$).

---
[1]This work is supported by the Russian Science Foundation under grant 21-11-00318.

Stepan L. Kuznetsov

Definition 3 *A given Kleene algebra $\mathcal{A}$ is commutative, if so is its multiplication operation: $a \cdot b = b \cdot a$ for all $a, b \in A$.*

Considering only *-continuous Kleene algebras looks rather natural, but, as we shall see below, this restriction of the class of algebras invokes raising of algorithmic complexity. As for commutativity, while most natural examples of Kleene algebras are non-commutative, the importance of the commutative case was noticed by Pratt (1991), from the point of view of modelling computation processes. Namely, commutative multiplication operation corresponds to parallel composition, as opposed to the sequential one.

Let us accurately formulate the logical language and define the problem whose complexity we are going to estimate. Formulae of this calculus are built from variables $(p, q, r, \ldots;$ the set of variables is denoted by Var) and constants 0 and 1 using the operations of Kleene algebra: $+$, $\cdot$, and *. Derivable *statements* are of the form $A \to B$, where A and B are formulae. Here $\to$ stands for $\preceq$, and we shall further prefer this notation.

Definition 4 *An interpretation of formulae on a given Kleene algebra $\mathcal{A}$ is defined by an interpreting function $v \colon \text{Var} \to A$, which is then propagated to the valuation function $\bar{v}$ for arbitrary formulae: $\bar{v}(p) = v(p)$ for $p \in \text{Var}$, $\bar{v}(0) = 0$, $\bar{v}(1) = 1$, $\bar{v}(A+B) = \bar{v}(A) + \bar{v}(B)$, $\bar{v}(A \cdot B) = \bar{v}(A) \cdot \bar{v}(B)$, $\bar{v}(A^*) = (\bar{v}(A))^*$. A statement $A \to B$ is true on $\mathcal{A}$ under interpretation v, if $\bar{v}(A) \preceq \bar{v}(B)$.*

Definition 5 *Let $\mathcal{H}$ be a set of statements. The set of statements $\mathcal{H}$ entails a statement $A \to B$ on a given class of Kleene algebras (e.g., commutative *-continuous Kleene algebras), if for any algebra $\mathcal{A}$ and any interpretation v on this algebra the following holds: if each statement from $\mathcal{H}$ is true on $\mathcal{A}$ under v, then so is $A \to B$.*

The *reasoning from hypotheses* problem for a given class of Kleene algebras is formulated as follows: *given a **finite** set of statements $\mathcal{H}$ and a statement $A \to B$, determine, whether $\mathcal{H}$ entails $A \to B$.*

From the point of view of complexity, the reasoning from hypotheses problem is interesting only in the *-continuous case. Otherwise, when the Kleene star is defined by a fixpoint condition (Definition 1.3), there is a trivial Σ_1^0 upper bound, because reasoning from hypotheses in this case can be axiomatised by a finitary calculus. On the other hand, the corresponding lower bound (Σ_1^0-hardness) holds even without Kleene star. Namely, in the non-commutative situation reasoning from hypotheses, without any occurrences of *, absorbs reasoning in semi-Thue systems. In the commutative situation, an analysis of the undecidability proof for commutative propositional linear logic by Lincoln, Mitchell, Scedrov, and Shankar (1992) shows that is actually proves Σ_1^0-hardness of reasoning from hypotheses in commutative idempotent semirings, in the language of $\cdot$ and $+$.

In the *-continuous situation, however, the complexity landscape is more interesting. Complexity of reasoning in (generally non-commutative) *-continuous Kleene algebras was studied by Kozen (2002). Kozen showed that in general it is Π_1^1-complete (!), while in the special case where $\mathcal{H}$ does not include * (while the goal sequent $A \to B$ could) it is Π_2^0-complete.

In this paper, we provide the commutative counterpart of Kozen's second result.

Theorem 1 *Reasoning from *-free hypotheses in *-continuous commutative Kleene algebras is Π_2^0-complete.*

The upper bound is easier and actually reduces to Kozen's non-commutative situation. Indeed, given (in the commutative setting) a set of *-free hypotheses $\mathcal{H}$ and a statement $A \to B$, let $p_1, \ldots, p_n$ be all variables used in $\mathcal{H} \cup \{A \to B\}$. Then let $\mathcal{H}' = \mathcal{H} \cup \{p_i \cdot p_j \to p_j \cdot p_i \mid 1 \leq i, j \leq n\}$. We claim that $\mathcal{H}$ entails $A \to B$ on commutative *-continuous Kleene algebras if and only if $\mathcal{H}'$ entails $A \to B$ on arbitrary *-continuous Kleene algebras. The "if" direction is trivial, since the new hypotheses are valid on all commutative algebras. For the "only if" direction, let $\mathcal{A}$ be a *-continuous Kleene algebra and v be an interpretation on it, such that all sequents from $\mathcal{H}'$ are true under this interpretation. Take the subalgebra of $\mathcal{A}$ generated by $v(p_1), \ldots, v(p_n)$. This subalgebra is commutative, since for generators it is explicitly postulated by $\mathcal{H}'$, and then commutativity propagates through Kleene algebra operations. Also, in this subalgebra, under interpretation v, all statements from $\mathcal{H}$ are true. Therefore, so is $A \to B$. It remains to notice that valuations of A and B are the same in the subalgebra and in $\mathcal{A}$ itself.

For the more interesting lower bound, we shall use a syntactic approach,[2] that is, reformulate entailment from hypotheses on *-continuous commutative Kleene algebras by derivability in an infinitary calculus denoted by **CommKA**$_\omega$.

The **CommKA**$_\omega$ calculus includes the following axioms:

$$A \to A \qquad 0 \cdot A \leftrightarrow 0 \qquad 1 \cdot A \leftrightarrow A$$

---

[2] Unlike Kozen (2002), whose approach is semantical.

$$A \to A + B \qquad B \to A + B \qquad (A + B) \cdot C \to (A \cdot C) + (B \cdot C)$$

$$(A \cdot B) \cdot C \leftrightarrow A \cdot (B \cdot C) \qquad A \cdot B \leftrightarrow B \cdot A \qquad A^n \to A^*$$

(as usual, $A \leftrightarrow B$ stands for two statements: $A \to B$ and $B \to A$) and the following rules of inference:

$$\frac{A \to B}{A \cdot C \to B \cdot C} \qquad \frac{A \to B \quad B \to C}{A \to C}$$

$$\frac{A \to C \quad B \to C}{A + B \to C} \qquad \frac{\left(A^n \cdot C \to D\right)_{n=0}^{\infty}}{A^* \cdot C \to D}$$

The last rule here is an ω-rule.

We say that a statement $A \to B$ is derivable in **CommKA**$_\omega$ from a set of statements $\mathcal{H}$, if $A \to B$ belongs to the smallest set which includes $\mathcal{H}$, includes all axioms of **CommKA**$_\omega$, and is closed under inference rules. For **CommKA**$_\omega$, the standard strong soundness-and-completeness theorem holds:

Proposition 1 *A set of statements $\mathcal{H}$ entails a statement $A \to B$ on commutative *-continuous Kleene algebras if and only if $A \to B$ is derivable from the set of hypotheses $\mathcal{H}$ in* **CommKA**$_\omega$.

Proof. Axioms and rules of **CommKA**$_\omega$ are valid on all commutative *-continuous Kleene algebras. This validates the "if" part. For the "only if" direction, we proceed by the standard Lindenbaum – Tarski construction. Let $A \approx_{\mathcal{H}} B$ mean that $A \to B$ and $B \to A$ are derivable in **CommKA**$_\omega$ from $\mathcal{H}$. This is clearly an equivalence relation. We construct a Kleene algebra on the factor-set of the set of formulae by this equivalence relation, i.e., the set of equivalence classes of formulae (the equivalence class of formula A is denoted by $[A]_{\approx_{\mathcal{H}}}$). On these equivalence classes, operations are defined in a standard way, say, $[A]_{\approx_{\mathcal{H}}} \cdot [B]_{\approx_{\mathcal{H}}} = [A \cdot B]_{\approx_{\mathcal{H}}}$. Also let $[A]_{\approx_{\mathcal{H}}} \preceq [B]_{\approx_{\mathcal{H}}}$ if $A \to B$ is derivable from $\mathcal{H}$ in **CommKA**$_\omega$. Correctness is checked in a routine way. Another routine check shows that the algebra constructed in this way is a commutative *-continuous Kleene algebra. Now suppose that a statement $A \to B$ is true on all algebras from this class. Then this statement is true, in particular, on the algebra we have just constructed, with the following interpretation of variables: $v(p) = [p]_{\approx_{\mathcal{H}}}$. Since, due to our definitions, $\bar{v}(A) = [A]_{\approx_{\mathcal{H}}}$ and similarly for B, we have $[A]_{\approx_{\mathcal{H}}} \preceq [B]_{\approx_{\mathcal{H}}}$, which means that $A \to B$ is derivable from $\mathcal{H}$ in **CommKA**$_\omega$. □

2 Internalising hypotheses using the exponential

In order to benefit even more from syntactic methods, we shall introduce a calculus which allows internalising hypotheses by a version of deduction theorem. This will be made possible by the exponential modality, !, taken from linear logic (Girard, 1987). The key feature of considering such a calculus is cut elimination. Besides operations of Kleene algebra and the exponential, this calculus will also include linear implication, $\multimap$ (which may also be regarded as division, or residuation). Thus, this system is an exponential extension of commutative infinitary action logic (Kuznetsov, 2020), more precisely, the fragment without additive conjunction. We denote it by **!CommACT**$_\omega$. This system is a commutative version of an exponential extension of infinitary action logic by Buszkowski and Palka (2008). We present **!CommACT**$_\omega$ in the form of sequent calculus.

Formulae of **!CommACT**$_\omega$ are built from a countable set of variables $\{p, q, r, \ldots\}$ and constants 0 and 1 using three binary operations: $+, \cdot, \multimap$ and two unary ones: $*$ and !. Formulae are denoted by capital Latin letters. Capital Greek letters stand for multisets of formulae. Sequents of **!CommACT**$_\omega$ are expressions of the form $\Pi \to B$. We use comma for both multiset union: Γ, Δ and for adding a formula to a multiset: Γ, A (if A were already in Γ, its multiplicity increases by one). A more formalistic notation would have been $\Gamma \uplus \Delta$ and $\Gamma \uplus \{A\}$ respectively. The notation A^n means $\underbrace{A, \ldots, A}_{n \text{ times}}$.

Axioms and inference rules of **!CommACT**$_\omega$ are as follows:

$$\frac{}{A \to A} \, Id \qquad \frac{}{\Gamma, 0 \to B} \, 0L \qquad \frac{\Gamma \to B}{\Gamma, 1 \to B} \, 1L \qquad \frac{}{\to 1} \, 1R$$

$$\frac{\Gamma, A \to C \quad \Gamma, B \to C}{\Gamma, A+B \to C} \, +L \qquad \frac{\Pi \to A}{\Pi \to A+B} \quad \frac{\Pi \to B}{\Pi \to A+B} \, +R$$

$$\frac{\Gamma, A, B \to C}{\Gamma, A \cdot B \to C} \, \cdot L \qquad \frac{\Gamma \to A \quad \Delta \to B}{\Gamma, \Delta \to A \cdot B} \, \cdot R$$

$$\frac{\Pi \to A \quad \Gamma, B \to C}{\Gamma, \Pi, A \multimap B \to C} \, \multimap L \qquad \frac{A, \Pi \to B}{\Pi \to A \multimap B} \, \multimap R$$

$$\frac{(\Gamma, A^n \to C)_{n=0}^{\infty}}{\Gamma, A^* \to C} \, *L \qquad \frac{\Pi_1 \to A \quad \ldots \quad \Pi_n \to A}{\Pi_1, \ldots, \Pi_n \to A^*} \, *R, \, n \geq 0$$

$$\frac{\Gamma, A \to C}{\Gamma, !A \to C} \, !L \qquad \frac{!A_1, \ldots, !A_n \to B}{!A_1, \ldots, !A_n \to !B} \, !R$$

$$\frac{\Gamma, !A, !A \to C}{\Gamma, !A \to C} \, !C \qquad \frac{\Gamma \to C}{\Gamma, !A \to C} \, !W$$

$$\frac{\Pi \to A \quad \Gamma, A \to C}{\Gamma, \Pi \to C} \, Cut$$

The calculus **!CommACT**$_\omega$ admits the following form of modalised deduction theorem (cf. Kanovich, Kuznetsov, Nigam, & Scedrov, 2019):

Theorem 2 *A sequent* $\Pi \to C$ *is derivable in* **!CommACT**$_\omega$ *with a finite number of sequents* $A_1 \to B_1, \ldots, A_n \to B_n$ *added as hypotheses (non-logical axioms) if and only if the sequent* $!(A_1 \multimap B_1), \ldots, !(A_n \multimap B_n), \Pi \to C$ *is derivable in* **!CommACT**$_\omega$ *without hypotheses.*

Proof. Let $!\Phi = !(A_1 \multimap B_1), \ldots, !(A_n \multimap B_n)$. For the "only if" part, we take the derivation of $\Pi \to C$ from hypotheses and append $!\Phi$ to the left-hand side of each sequent in this derivation. This transforms axioms Id, $0L$, and $1R$ to sequents derivable from the corresponding axioms using $!W$ (for formulae in $!\Phi$). One-premise rules remain valid (in particular, $!R$, since the formulae added are of the form $!F$). For two-premise rules, the conclusion receives two copies of $!\Phi$, which we contract using $!C$. Finally, hypotheses of the form $A_i \to B_i$ become derivable sequents:

$$\frac{\dfrac{\dfrac{A_i \to A_i \quad B_i \to B_i}{A_i \multimap B_i, A_i \to B_i} \multimap L}{!(A_i \multimap B_i), A_i \to B_i} \, !L}{!\Phi, A_i \to B_i} \, !W, n-1 \text{ times}$$

For the "if" part, given $!\Phi, \Pi \to C$, we derive $\Pi \to C$ using Cut and our hypotheses, in the following way:

$$\frac{\dfrac{\dfrac{A_1 \to B_1}{\to A_1 \multimap B_1} \multimap R}{\to !(A_1 \multimap B_1)} \, !R \quad \ldots \quad \to !(A_n \multimap B_n) \quad !\Phi, \Pi \to C}{\Pi \to C} \, Cut, n \text{ times}$$

$\square$

It is easy to see that axioms and rules of **CommKA**$_\omega$ are derivable in **!CommACT**$_\omega$. In the view of Proposition 1 this yields the following

Corollary 1 *If a set of statements $\mathcal{H}$ entails a statement $A \to B$ on commutative *-continuous Kleene algebras, then $A \to B$ is derivable from $\mathcal{H}$ in* **!CommACT**$_\omega$.

Actually, the opposite direction here also holds, but this is not that easy, since Theorem 2 gives only derivability in **!CommACT**$_\omega$, not in **CommKA**$_\omega$. Fortunately, we shall not need this opposite direction, so we omit its proof.

The key feature of **!CommACT**$_\omega$, which makes it more useful than deriving from hypotheses in **CommKA**$_\omega$, is cut elimination:

Theorem 3 *Any sequent derivable in* **!CommACT**$_\omega$ *(without hypotheses) is derivable without using the Cut rule.*

The proof of cut elimination is a straightforward commutative modification of cut elimination for non-commutative action logic with exponentiation (Kuznetsov & Speranski, 2022), so we omit it.

A standard usage of Cut is for establishing invertibility of certain inference rules:

Proposition 2 *Rules $+L$, $\cdot L$, and $*L$ are invertible, that is, derivability of the conclusion of such a rule entails derivability of all its premises.*

Proof. By Cut with $A \to A+B$ or $B \to A+B$; $A, B \to A \cdot B$; $A^n \to A^*$, respectively. □

3 Encoding totality for counter machines

Following Kozen (2002), we encode the Π^0_2-complete problem of *totality:* given a machine, determine whether it halts on all inputs. In the commutative setting, however, it is hardly possible to encode Turing machines, since we cannot maintain the order of letters on the tape. Therefore, we opt for counter machines (Minsky, 1961). The same idea is used in the proof of undecidability of commutative linear logic (Lincoln et al., 1992) and commutative infinitary action logic (Kuznetsov, 2020). The price of the convenience of counter machines is that we now need additive disjunction (join, which we denote by $+$); fortunately, in Kleene algebras we have it.

For our purposes, three counters will be enough, so let $\mathcal{R} = \{\mathsf{a}, \mathsf{b}, \mathsf{c}\}$ be the set of counters.

Stepan L. Kuznetsov

Definition 6 *A deterministic 3-counter machine $\mathfrak{M}$ consists of a finite set of states $\mathcal{Q}$, designated initial and final states $q_I, q_F \in \mathcal{Q}$, and a finite set of instructions $\mathcal{I}$. Each instruction is of one of the following two forms and with the following informal meaning (here $r \in \mathcal{R}$):*

$\text{INC}(p, r, q)$	*from state p, increase counter r by 1 and move to q;*
$\text{JZDEC}(p, r, q_0, q_1)$	*from state p, if r is zero, move to q_0, otherwise decrease r by 1 and move to q_1.*

For each $p \in \mathcal{Q} - \{q_F\}$ there exists exactly one instruction in $\mathcal{I}$ with p as its first parameter. For $p = q_F$ there is no such instruction.

Definition 7 *A configuration of a 3-counter machine is a quadruple of the form $\langle p, a, b, c \rangle$, where $p \in \mathcal{Q}$ and a, b, c are natural numbers, meaning the values of the counters. A configuration $\mathbf{c}_2$ is the immediate successor of another configuration $\mathbf{c}_1$, if $\mathbf{c}_2$ is obtained from $\mathbf{c}_1$ by applying an instruction of $\mathfrak{M}$ (due to determinism, the immediate successor is unique, if it exists). Machine $\mathfrak{M}$ reaches configuration $\mathbf{c}_2$ from configuration $\mathbf{c}_1$, if there is a chain of immediate successors connecting $\mathbf{c}_1$ to $\mathbf{c}_2$.*

Definition 8 *The partial function $f_\mathfrak{M} \colon \mathbb{N} \to \mathbb{N}$ computed by $\mathfrak{M}$ is defined as follows:*

- *$f(a) = b$, if $\mathfrak{M}$ reaches $\langle q_F, b, 0, 0 \rangle$ from $\langle q_I, a, 0, 0 \rangle$;*

- *$f(a)$ is undefined otherwise.*

Three counters are sufficient for Turing completeness in the sense of this definition:[3]

Proposition 3 (Schroeppel, 1972) *A partial function $f \colon \mathbb{N} \to \mathbb{N}$ is computable by a Turing machine $\mathfrak{T}$ if and only if it is computable by a 3-counter machine $\mathfrak{M}$. Moreover, the translation from $\mathfrak{T}$ to $\mathfrak{M}$ is itself computable.*

Corollary 2 *The following totality problem for 3-counter machines is Π_2^0-complete: given a 3-counter machine $\mathfrak{M}$, determine whether $f_\mathfrak{M}$ is defined on all natural numbers. (Such machines will be called total.)*

---

[3]Two counters also give a Turing-complete computational model (Minsky, 1961), but in a special sense: natural numbers serving as input and output should be encoded as 2^n; the function $n \mapsto 2^n$ itself is not computable by a 2-counter machine (Schroeppel, 1972). Thus, computing with three counters is more convenient.

Definition 9 *Given a 3-counter machine $\mathfrak{M}$, we construct a set of *-free hypotheses $\mathcal{H}_\mathfrak{M}$ in the following way:*

- *for each rule of the form $\text{INC}(p, r, q)$, we add $p \to q \cdot r$;*
- *for each rule of the form $\text{JZDEC}(p, r, q_0, q_1)$, we add $p \cdot r \to q_1$ and $p \to q_0 + z_r$.*

(We suppose that $\mathcal{Q} \cup \{\mathsf{a}, \mathsf{b}, \mathsf{c}, z_\mathsf{a}, z_\mathsf{b}, z_\mathsf{c}\}$ is a subset of the set of variables.)

The main encoding statement is now as follows:

Theorem 4 *Machine $\mathfrak{M}$ is total if and only if $\mathcal{H}_\mathfrak{M}$ entails the following statement:*

$$q_I \cdot \mathsf{a}^* \to (q_F \cdot \mathsf{a}^*) + (z_\mathsf{a} \cdot \mathsf{b}^* \cdot \mathsf{c}^*) + (z_\mathsf{b} \cdot \mathsf{a}^* \cdot \mathsf{c}^*) + (z_\mathsf{c} \cdot \mathsf{a}^* \cdot \mathsf{b}^*)$$

*on *-continuous commutative Kleene algebras.*

Theorem 1 immediately follows from Theorem 4 and Corollary 2. The next two sections contain the proof of Theorem 4. In what follows, we denote $(q_F \cdot \mathsf{a}^*) + (z_\mathsf{a} \cdot \mathsf{b}^* \cdot \mathsf{c}^*) + (z_\mathsf{b} \cdot \mathsf{a}^* \cdot \mathsf{c}^*) + (z_\mathsf{c} \cdot \mathsf{a}^* \cdot \mathsf{b}^*)$ by D.

4 From computations to derivations

We start with proving the easier "only if" direction of Theorem 4. In the view of Proposition 1, we shall establish syntactic derivability of the sequent $q_I \cdot \mathsf{a}^* \to D$ from $\mathcal{H}_\mathfrak{M}$ instead of semantic entailment. We shall take an arbitrary input value x for $f_\mathfrak{M}$ and derive $q_I \cdot \mathsf{a}^x \to D$. Let us prove the following more general lemma:

Lemma 1 *If $\mathfrak{M}$ reaches a configuration of the form $\langle q_F, y, 0, 0\rangle$ from configuration $\langle p, a, b, c\rangle$, then the following statement*

$$p \cdot \mathsf{a}^a \cdot \mathsf{b}^b \cdot \mathsf{c}^c \to D$$

is derivable from $\mathcal{H}_\mathfrak{M}$ in $\mathbf{CommKA}_\omega$.

For $p = q_I$, $a = x$, $b = c = 0$, this lemma gives exactly $q_I \cdot \mathsf{a}^x \to D$. The reachability condition follows from totality of $\mathfrak{M}$. Since x was taken arbitrary, we establish the original sequent $q_I \cdot \mathsf{a}^* \to D$.

Proof. If $p = q_F$, then by our definitions the machine should already be in a valid final configuration. This means $b = c = 0$, and the sequent in question is derived by transitivity from $q_F \cdot \mathsf{a}^a \to q_F \cdot \mathsf{a}^*$ and $q_F \cdot \mathsf{a}^* \to D$. The former is an axiom of **CommKA**$_\omega$ and the latter is easily derivable.

If $p \ne q_F$, consider the instruction applied immediately (since $\mathfrak{M}$ is deterministic, this instruction is unique). Without loss of generality we may suppose that r, the active counter in the instruction, is a.

For $\mathrm{INC}(p, \mathsf{a}, q)$, the process is straightforward. We derive $p \cdot \mathsf{a}^a \cdot \mathsf{b}^b \cdot \mathsf{c}^c \to q \cdot \mathsf{a}^{a+1} \cdot \mathsf{b}^b \cdot \mathsf{c}^c$ and then proceed by induction hypotheses and transitivity. This statement, in its turn, follows by monotonicity from $p \to q \cdot \mathsf{a}$, which belongs to $\mathcal{H}_\mathfrak{M}$. The same happens for $\mathrm{JZDEC}(p, \mathsf{a}, q_0, q_1)$ when $a > 0$: here we use $p \cdot \mathsf{a} \to q_1$, which is also in $\mathcal{H}_\mathfrak{M}$.

The case of $\mathrm{JZDEC}(p, \mathsf{a}, q_0, q_1)$ with $a = 0$ is a bit more involved. Here we use $p \to q_0 + z_\mathsf{a}$. By monotonicity we get $p \cdot \mathsf{b}^b \cdot \mathsf{c}^c \to (q_0 + z_\mathsf{a}) \cdot \mathsf{b}^b \cdot \mathsf{c}^c$ (recall that there is no a^a, since $a = 0$), and by distributivity of $\cdot$ over $+$ we have $p \cdot \mathsf{b}^b \cdot \mathsf{c}^c \to q_0 \cdot \mathsf{b}^b \cdot \mathsf{c}^c + z_\mathsf{a} \cdot \mathsf{b}^b \cdot \mathsf{c}^c$. By induction hypothesis we have $q_0 \cdot \mathsf{b}^b \cdot \mathsf{c}^c \to D$; we also can derive $z_\mathsf{a} \cdot \mathsf{b}^b \cdot \mathsf{c}^c \to z_\mathsf{a} \cdot \mathsf{b}^* \cdot \mathsf{c}^*$, which yields $z_\mathsf{a} \cdot \mathsf{b}^b \cdot \mathsf{c}^c \to D$. Thus, we get $q_0 \cdot \mathsf{b}^b \cdot \mathsf{c}^c + z_\mathsf{a} \cdot \mathsf{b}^b \cdot \mathsf{c}^c \to D$, and by transitivity $p \cdot \mathsf{b}^b \cdot \mathsf{c}^c \to D$. $\square$

5 From derivations to computations

Now let us prove the "if" direction of Theorem 4. Recall that $D = (q_F \cdot \mathsf{a}^*) + (z_\mathsf{a} \cdot \mathsf{b}^* \cdot \mathsf{c}^*) + (z_\mathsf{b} \cdot \mathsf{a}^* \cdot \mathsf{c}^*) + (z_\mathsf{c} \cdot \mathsf{a}^* \cdot \mathsf{b}^*)$ and let

$$!\mathcal{H}_\mathfrak{M} = \{\,!(A \multimap B) \mid (A \to B) \in \mathcal{H}_\mathfrak{M}\},$$

considered as a multiset where each formula has multiplicity 1. By Corollary 1, since $\mathcal{H}_\mathfrak{M}$ entails $q_I \cdot \mathsf{a}^* \to D$ on commutative *-continuous Kleene algebras, the sequent

$$!\mathcal{H}_\mathfrak{M}, q_I \cdot \mathsf{a}^* \to D$$

is derivable in **!CommACT**$_\omega$. Fix an arbitrary natural number x. By invertibility of $\cdot L$ and $*L$ (Proposition 2) we conclude that the following sequent is also derivable:

$$!\mathcal{H}_\mathfrak{M}, q_I, \mathsf{a}^x \to D.$$

Consider a cut-free proof of this sequent. The key observation is that this proof is finite: indeed, the only occurrences of $*$ are in D, and they are introduced by $*R$ (which is a finitary rule), not $*L$.

Let us erase all formulae of the form $!F$ from the proof. This trivialises rules $!C$ and $!W$ ($!R$ is never used), and $!L$ transforms into the following strange rule:

$$\frac{\Gamma, F \to C}{\Gamma \to C} \text{ , where } !F \in !\mathcal{H}_{\mathfrak{M}}.$$

Let us denote this rule by H. The only rules which can be applied in the proof are now $*R$, $+R$, $\cdot R$, $\multimap L$, $+L$, $\cdot L$, and H. Moreover, $+L$ and $\cdot R$ are invertible (Proposition 2), so we may suppose that they are applied immediately when possible.

Now we perform *disbalancing transformations,* which make the derivation exactly correspond to a counter machine computation. These transformations are actually the same as in Kuznetsov (2021), but in the commutative situation.

First, we notice that the $\multimap R$ rule is never applied, so right-hand sides are actually independent from the left-hand sides. This means that we may propagate the "right" rules, $*R$, $+R$, and $\cdot R$, to the top of the derivation tree. Now no "left" rule appears above a "right" one.

Second, we define the *main branch* of the derivation tree. The only two-premise rules are $\multimap L$, $+L$, and $\cdot R$. After the transformation performed above, the latter is applied only on the very top of the tree, and we stop the main branch when it reaches $\cdot R$. For $\multimap L$, the main branch goes to its right ("main") premise; for $+L$, which decomposes a formula of the form $q_0 + z_r$, it goes to the branch with q_0.

Now we notice that using the $+L$ rule is impossible outside the main branch. Indeed, since this rule should introduce $q_0 + z_r$, one of its premises should include z_r in its left-hand side. This instance of z_r goes upwards to one of the axioms, therefore, there should be a complementary z_r in the right-hand side. This could happen, however, only in D, which keeps in the main branch. (When the main branch stops and D starts being decomposed, we are allowed to use only right rules, so $+L$ also could not appear above the main branch.)

Thus, the "left" rules which could be used outside the main branch are the following ones: $\cdot L$, $\multimap L$, and H. Now we are at the core of the disbalancing procedure. Due to the shape of formulae of the form $A \multimap B$, which all come from $!\mathcal{H}_{\mathfrak{M}}$, left premises of $\multimap L$ on the main branch are of the form $\Pi \to p$ or $\Pi \to p \cdot r$. We shall rebuild our derivation tree, so that these premises trivialise, i.e., Π will be just p or p, r respectively.

Suppose Π is more complex, which means that some "left" rules were applied above the corresponding left premise, outside the main branch. We perform the following transformations, which move these rules to the main branch.

For $\cdot L$,

$$\cfrac{\cfrac{\Pi, E, F \to A}{\Pi, E \cdot F \to A} \cdot L \quad \Gamma, B \to D}{\Gamma, \Pi, E \cdot F, A \multimap B \to D} \multimap L$$

transforms into

$$\cfrac{\cfrac{\Pi, E, F \to A \quad \Gamma, B \to D}{\Gamma, \Pi, E, F, A \multimap B \to D} \multimap L}{\Gamma, \Pi, E \cdot F, A \multimap B \to D} \cdot L$$

For $\multimap L$,

$$\cfrac{\cfrac{\Pi_2 \to E \quad \Pi_1, F \to A}{\Pi_1, \Pi_2, E \multimap F \to A} \multimap L \quad \Gamma, B \to D}{\Gamma, \Pi_1, \Pi_2, E \multimap F, A \multimap B \to D} \multimap L$$

transforms into

$$\cfrac{\Pi_2 \to E \quad \cfrac{\Pi_1, F \to A \quad \Gamma, B \to D}{\Gamma, \Pi_1, F, A \multimap B \to D} \multimap L}{\Gamma, \Pi_1, \Pi_2, E \multimap F, A \multimap B \to D} \multimap L$$

For H,

$$\cfrac{\cfrac{\Pi, F \to A}{\Pi \to A} H \quad \Gamma, B \to D}{\Gamma, \Pi, A \multimap B \to D} \multimap L$$

transforms into

$$\cfrac{\cfrac{\Pi, F \to A \quad \Gamma, B \to D}{\Gamma, \Pi, F, A \multimap B \to D} \multimap L}{\Gamma, \Pi, A \multimap B \to D} H$$

After applying all possible transformations of these forms (recall that our proof tree is finite, so it is a finite process), we get a *disbalanced* proof tree with the following property: for each application of $\multimap L$, its left premise is $p \to p$ or $p, r \to p \cdot r$.

Finally, we may suppose that the formula F in the H rule gets decomposed immediately after above the H rule application (otherwise we could keep it in the hidden $!F$ form until it is needed).

Reasoning in Commutative Kleene Algebras

Having all these transformations performed, we now start the analysis of our derivation. As for the "from computations to derivations" direction, in order to proceed by induction we consider the more general sequent,

$$p, \mathsf{a}^a, \mathsf{b}^b, \mathsf{c}^c \to D.$$

Our claim which we prove by induction on the derivation of this sequent, is as follows: if this sequent is derivable, then $\mathfrak{M}$ terminates at a configuration of the form $\langle q_F, x, 0, 0 \rangle$ when started from $\langle p, a, b, c \rangle$.

If this sequent is derived using only "right" rules, we have $p = q_F$, which means that the computation successfully terminates. The form of D guarantees that $b = c = 0$.

Otherwise, the lowermost rule should be H, and immediately above it we have $\multimap L$ which decomposes its active formula F. Consider three cases depending of the possible forms of F (recall that $F = A \multimap B$, where $(A \to B) \in \mathcal{H}_\mathfrak{M}$).

Case 1: $F = p \multimap q \cdot r$, which means that $\text{INC}(p, r, q)$ is an instruction of $\mathfrak{M}$. The rule applied next is $\multimap L$, and since our derivation is disbalanced, its left premise is $p \to p$. Let $r = \mathsf{a}$ (without loss of generality), then the right premise is

$$q \cdot \mathsf{a}, \mathsf{a}^a, \mathsf{b}^b, \mathsf{c}^c \to D.$$

Immediately applying the invertible rule $\cdot L$, we get

$$q, \mathsf{a}^{a+1}, \mathsf{b}^b, \mathsf{c}^c \to D.$$

Now, by induction hypothesis, $\mathfrak{M}$ reaches its final configuration from $\langle q, a + 1, b, c \rangle$, and therefore from $\langle p, a, b, c \rangle$, $\text{INC}(p, \mathsf{a}, q)$ being the first step.

Case 2: $F = (p \cdot r) \multimap q_1$. This case is considered in the same fashion as Case 1, with $p, r \to p \cdot r$ being the left premise and $\text{JZDEC}(p, r, q_0, q_1)$ being the rule applied, with the value of r being non-zero.

Case 3: $F = p \multimap (q_0 + z_r)$. This is the most interesting case. Again, let $r = \mathsf{a}$. Now our left premise is $p \to p$, the right one is

$$q_0 + z_\mathsf{a}, \mathsf{a}^a, \mathsf{b}^b, \mathsf{c}^c \to D.$$

We immediately apply $+L$ (invertible rule), getting two premises:

$$q_0, \mathsf{a}^a, \mathsf{b}^b, \mathsf{c}^c \to D \quad \text{and} \quad z_\mathsf{a}, \mathsf{a}^a, \mathsf{b}^b, \mathsf{c}^c \to D.$$

Consider the derivation of the second one and suppose it involves using the H rule, followed by the corresponding $\multimap L$. This adds a variable $q \in \mathcal{Q}$

into the left-hand side of this sequent, and when tracing upwards, following the sequents with z_a, such an element is persistent (the concrete element of $\mathcal{Q}$ changes, but we always have *some* $q' \in \mathcal{Q}$). Thus, on the top we get a sequent with *both* $q \in \mathcal{Q}$ and z_a in its left-hand side. On the right, we have either D, or p, or $p \cdot r$. It is easy to see that such a sequent cannot be derived using "right" rules. Contradiction.

Therefore, the sequent $z_\mathsf{a}, \mathsf{a}^a, \mathsf{b}^b, \mathsf{c}^c \to D$ is derived immediately using "right" rules, which yields, due to the form of D, $a = 0$. This means that we are in the "zero" case of JZDEC, and the step is from $\langle p, 0, b, c \rangle$ to $\langle q_0, 0, b, c \rangle$. The latter configuration exactly correspond to our first sequent, $q_0, \mathsf{b}^b, \mathsf{c}^c \to D$, and we proceed by induction.

In particular, our argument yields the following: if $!\mathcal{H}_\mathfrak{M}, q_I, \mathsf{a}^x \to D$ is derivable (here we have restored the hidden !-formulae), then $\mathfrak{M}$ reaches a final configuration starting from $\langle q_I, x, 0, 0 \rangle$, in other words, $f_\mathfrak{M}(x)$ is defined. Since this sequent was derivable for arbitrary x (this follows from derivability of $!\mathcal{H}_\mathfrak{M}, q_I, \mathsf{a}^* \to D$), we get totality of $\mathfrak{M}$.

This finishes the proof of Theorem 4 and therefore of Theorem 1.

6 Conclusion

In this paper, we have shown that reasoning from *-free hypotheses in commutative *-continuous Kleene algebras is Π_2^0-complete. This serves as a commutative counterpart to a result by Kozen (2002). In order to facilitate the proof of the lower bound, we have used an intermediate calculus **!CommACT**$_\omega$, commutative infinitary action logic with exponentiation. The complexity of **!CommACT**$_\omega$, however, is probably much higher than Π_2^0. This system can encode reasoning from arbitrary hypotheses, and we conjecture that this is Π_1^1-complete. Proving this would give a commutative counterpart for another result of Kozen (2002).

References

Buszkowski, W., & Palka, E. (2008). Infinitary action logic: Complexity, models and grammars. *Studia Logica, 89*(1), 1–18.

Girard, J.-Y. (1987). Linear logic. *Theoretical Computer Science, 50*(1), 1–101.

Kanovich, M., Kuznetsov, S., Nigam, V., & Scedrov, A. (2019). Subexponentials in non-commutative linear logic. *Mathematical Structures in Computer Science, 29*(8), 1217–1249.

Kleene, S. C. (1956). Representation of events in nerve nets and finite automata. In *Automata Studies* (pp. 3–41). Princeton: Princeton University Press.

Kozen, D. (2002). On the complexity of reasoning in Kleene algebra. *Information and Computation, 179*(2), 152–162.

Kuznetsov, S. (2020). Complexity of commutative infinitary action logic. In M. A. Martins & I. Sedlár (Eds.), *DaLí 2020: Dynamic Logic, New Trends and Applications* (pp. 155–169). Cham: Springer.

Kuznetsov, S. L. (2021). Kleene star, subexponentials without contraction, and infinite computations. *Siberian Electronic Mathematical Reports, 18*(2), 905–922.

Kuznetsov, S. L., & Speranski, S. O. (2022). Infinitary action logic with exponentiation. *Annals of Pure and Applied Logic, 173*(2). (Published online, article no. 103057)

Lincoln, P., Mitchell, J., Scedrov, A., & Shankar, N. (1992). Decision problems for propositional linear logic. *Annals of Pure and Applied Logic, 56*(1–3), 239–311.

Minsky, M. L. (1961). Recursive unsolvability of Post's problem of "Tag" and other topics in theory of Turing machines. *Annals of Mathematics, 74*(3), 437–455.

Pratt, V. (1991). Action logic and pure induction. In *JELIA 1990: Logics in AI* (Vol. 478, pp. 97–120). Berlin, Heidelberg: Springer-Verlag.

Schroeppel, R. (1972). *A two counter machine cannot calculate 2^N*. Massachusets Institute of Technology A.I. Laboratory, Artificial Intelligence Memo #257.

Stepan L. Kuznetsov
Russian Academy of Sciences, Steklov Mathematical Institute
Russia
E-mail: sk@mi-ras.ru

Hyperdoctrine Semantics: An Invitation

SHAY LOGAN AND GRAHAM LEACH-KROUSE[1]

Abstract: Categorial logic, as its name suggests, applies the techniques and machinery of category theory to topics traditionally classified as part of logic. We claim that these tools deserve attention from a greater range of philosophers than just the mathematical logicians. We support this claim with an example. In this paper we show how one particular tool from categorial logic—hyperdoctrines—suggests interesting metaphysics. Hyperdoctrines can provide semantics for quantified languages, but this account of quantification suggests a metaphysical picture quite different from the one suggested by standard model-theoretic semantics.

Keywords: hyperdoctrines, categorical logic, metaphysics, semantics

In this paper, we wish to suggest that a tool from category theory, and in particular categorial logic—the theory of hyperdoctrines—is of metaphysical interest. It presents an alternative to a viewpoint that has become entrenched (at least in some circles) to the point of invisibility. The first three sections of our paper are a crash-course in hyperdoctrine semantics for classical first-order logic. The final section argues that a focus on first-order model theory has distorted many philosopher's metaphysical theorizing, and uses the results of the first three sections to sketch an alternative.

1 Language and logic

We call the language we work with throughout this paper $\mathcal{L}$. Each well-formed expression in $\mathcal{L}$ has the form '$\phi \mid X$' with ϕ a sequence of symbols called the *untyped part* of the expression and X a set of variables called the *typing part* of the expression. Philosophically, we understand the typing part of an $\mathcal{L}$-expression to specify something like the 'dimensions' along which the untyped part is taken to be incomplete.

---

[1] Many thanks to the audience at Logica, as well as to Andrew Tedder, Teresa Kouri, Eileen Nutting, Blane Worley, and an anonymous referee, who helped us sharpen and refine the ideas that follow.

The typing part can make a difference to the meaning of an expression even if not all of its variables appear in the untyped part. As an aid to understanding, consider a polynomial like $x - y$. The set of points at which this polynomial is zero can be viewed as a line in a two dimensional space if the only variables under consideration are x and y. But it can also be viewed as a plane in a three dimensional space when the variable z is in play. In $\mathcal{L}$, context matters: the zero locus of $x - y$ in the $\{x, y\}$-context is a line; the zero locus of $x - y$ in the $\{x, y, z\}$-context is a plane.

Formally, the terms of our language are constructed from a vocabulary consisting of variables $x_1, x_2, \ldots$, constants $c_1, c_2, \ldots$, the separator symbol '|', and the brackets '{' and '}', which we will tend to drop. We will regard a sequence τ of variables and constants with a shared context X as a single term $\tau \mid X$. If there are n terms in τ, we will say that τ is n-ary.

Definition 1 (Terms of $\mathcal{L}$) *If c is a constant, then $c \mid \varnothing$ is a unary term. If x is a variable, then $x \mid \{x\}$ is a unary term. If $\tau \mid X$ is an n-ary term and $\sigma \mid Y$ is an m-ary term, then $\tau \sigma \mid X \cup Y$ is an $n + m$-ary term. If $\tau \mid X$ is an n-ary term and y is a variable, then $\tau \mid X \cup \{y\}$ is an n-ary term.*

For each n we recognize n-adic predicates $R_n^1, R_n^2, \ldots$. For concreteness, we will recognize three connectives ($\neg$, $\wedge$, and $\rightarrow$), one quantifier ($\forall$), and take '$\exists$' and '$\vee$' to be defined. We specify the set of formulas as follows:

Definition 2 (Formulas of $\mathcal{L}$) *If R is an n-adic predicate and $\tau \mid X$ is an n-ary term, then $R\tau \mid X$ is a formula. If $\phi \mid X$ and $\psi \mid X$ are formulas, then so are $\neg \phi \mid X$, $(\phi \wedge \psi) \mid X$, and $(\phi \rightarrow \psi) \mid X$. If $\phi \mid X$ is a formula and y is a variable, then $\phi \mid X \cup \{y\}$ is a formula. Last, if $\phi \mid X$ is a formula and $x \in X$, then $\forall x \phi \mid X - \{x\}$ is a formula.*

We adopt the usual conventions regarding outermost parentheses and similar matters. To indicate substitutions, we declare that if τ, σ, and η are constants or variables, then $\tau(\sigma/\eta)$ is η if $\tau = \sigma$, and otherwise τ. We read this as "replace σ with η". If $\overline{\sigma}$ and $\overline{\eta}$ are sequences of constants or variables, then $(\overline{\sigma}/\overline{\eta})$ abbreviates a simultaneous replacement of each σ_i with η_i. We suppose that this is done in some way that avoids collision.

If ϕ is the untyped part of a formula, $\phi(\sigma/\eta)$ is the result of replacing each free occurrence of a constant or variable t in ϕ by an occurrence of $t(\sigma/\eta)$. If X is a set of variables, then $\overline{X}$ abbreviates the sequence of those variables taken in increasing order (by their subscript indices). If X and Y are sets of variables with $\text{card}(X) = \text{card}(Y)$ then we call a replacement of

Hyperdoctrine Semantics

the form $\phi(\overline{Y}/\overline{X})$ a change of variables. Finally, we say that $\phi(\overline{\sigma}/\overline{\eta})$ is a *proper* substitution instance of ϕ if each η_i is freely substitutable for σ_i in ϕ.

We will write **K** for the subset of $\mathcal{L}$ that is, apart from typing, plain-old classical logic. More to the point, we say that if ϕ is a theorem of classical logic and $\phi \mid X$ is well formed, then $\phi \mid X \in \mathbf{K}$. But our main interest in what follows is not actually in **K** itself, but in the notion of **K**-provability. For the latter, we restrict to the special case of single-premise provability.[2] We will write $\phi \mid X \vdash_\mathbf{K} \psi \mid Y$ to mean that $\psi \mid Y$ is **K**-provable from $\phi \mid X$, and we define this relation as follows:

Definition 3 $\phi \mid X \vdash_\mathbf{K} \psi \mid Y$ *iff there is a sequence of formulas* $\psi_1 \mid X_1, \psi_2 \mid X_2, \ldots, \psi_n \mid X_n$ *with* $\psi_n \mid X_n = \psi \mid Y$ *such that for all* $1 \leq i \leq n$, *either* $\psi_i \mid X_i = \phi \mid X$, *or for some* $j < i$, $X_j = X_i$ *and* $\psi_j \to \psi_i \mid X_j \in \mathbf{K}$, *or for some* $j < i$ *and* $k < i$, $X_i = X_j = X_k$ *and* $\psi_i = \psi_j \wedge \psi_k$, *or for some* $j < i$, $\psi_i = \psi_j(\overline{X_j}/\overline{X_i})$ *is a proper substitution instance of* ψ_j.

Note that we allow a proper change of variables in the course of a proof. Why? Consider $x = a \mid x$ and $y = a \mid y$.[3] $x = a$, regarded as incomplete only along the x-dimension, defines the same property (intuitively the property of being identical to a) as $y = a$ does when regarded as incomplete only along the y-dimension. So we ought to adopt a mechanism that lets us count $x = a \mid x$ as expressing the same thing as $y = a \mid y$. To accomplish this, it's clear we ought to adopt *some sort* of variable-substitution policy.

Not just any policy will do, though. Changes of variables in our technical sense are always *monotone*. To see why this must be, consider the formulas $x \leq y \mid x, y$ and $y \leq x \mid x, y$. Each of these formulas defines the less-than-or-equal-to relation. Yet we wouldn't want to regard them as equivalent—if they were equivalent, then conjoining them wouldn't give us anything new. But of course it does: $x \leq y \wedge y \leq x \mid x, y$ defines the identity relation, which neither $x \leq y \mid x, y$ nor $y \leq x \mid x, y$, taken on its own, does.

What goes wrong is that $x \leq y \mid x, y$ and $y \leq x \mid x, y$ are true of different sets of tuples. If the first is true of $1, 2$ (in that order), then the second is true of $2, 1$ (in *that* order). Now, recall that our variables come equipped with an ordering: the one imposed on them by their subscripts. This imposes an ordering on the dimensions of incompleteness of a formula expressed in

---

[2]The extension to multipremise provability is straightforward, provided all the premises are required to share a context.

[3]Neither of these is in fact a formula in $\mathcal{L}$. Don't get hung up on this; just pretend for a moment that we have a language with identity (and, in a moment, with inequalities).

those variables. Formulas can be regarded as being "true of" a sequence of things if the result of supplying the members of the sequence to the dimensions of incompleteness *in order* is true. That is why our extended notion of provability only allows variable changes that are *monotone*: monotone changes preserve the feature of tracking which tuples a formula is 'true of'.

Now that you know why we've defined **K**-provability the way we have, we end the section by noting a few important facts about this relation:

Lemma 1 *If $\phi \mid X \vdash_{\mathbf{K}} \psi \mid Y$, then $\phi(\overline{X}/\overline{Y}) \to \psi \mid Y \in \mathbf{K}$.*

Lemma 2 *If $\phi \mid X \vdash_{\mathbf{K}} \psi \mid Y$, then $\phi \mid X \cup \{z\} \vdash_{\mathbf{K}} \psi \mid Y \cup \{z\}$.*

It turns out that there is a natural way to view (equivalence classes of) terms of $\mathcal{L}$ as arrows in a category. Our goal at the moment is to very concretely describe this category.

To help keep concepts clearly delimited, we will write $\langle \tau \mid X \rangle$ for the arrow associated with the term $\tau \mid X$. As a preview of what's to come, we offer the following summary: the category we are constructing has for its objects the 'types' T_n, where n is a natural number. Each arrow $\langle \tau \mid X \rangle$ will have domain $T_{\text{card}(X)}$ and codomain $T_{\text{len}(\tau)}$. It follows that $\langle \tau_2 \mid X_2 \rangle \circ \langle \tau_1 \mid X_1 \rangle$ is defined just when $\text{len}(\tau_1) = \text{card}(X_2)$. Composition in $\mathcal{B}$ is just careful substitution. That is, the composition $\langle \tau_2 \mid X_2 \rangle \circ \langle \tau_1 \mid X_1 \rangle$ is formed by substituting the symbols constituting τ_1 for the variables occurring in X_2.

To say a bit more, it helps to first look at an example: consider $\langle xy \mid x, y, z \rangle : T_3 \longrightarrow T_2$ and $\langle avvw \mid v, w \rangle : T_2 \longrightarrow T_4$. The composition $\langle avvw \mid v, w \rangle \circ \langle xy \mid x, y, z \rangle$ should be an arrow $T_3 \longrightarrow T_4$. Here's how to make this happen: pair xy in the untyped part of $\langle xy \mid x, y, z \rangle$ with v, w in the typing part of $\langle avvw \mid v, w \rangle$. Use that pairing to replace the symbols of $avvw$, creating a new untyped part $axxy$ compatible with the typing part x, y, z. Or, in a picture, composition works like this in the case at hand:

$$\left\langle \begin{array}{c} a \\ v \leftarrow \mid -v \leftarrow \\ v \leftarrow' \mid -w \leftarrow \\ w \leftarrow' \end{array} \right\rangle \circ \left\langle \begin{array}{c} -x \leftarrow \\ -y \leftarrow \end{array} \right| \begin{array}{c} -x \\ -y \\ z \end{array} \right\rangle$$

Either way we describe it, the result is the same:

$$\langle avvw \mid v, w \rangle \circ \langle xy \mid x, y, z \rangle = \langle axxy \mid x, y, z \rangle$$

More generally:

Hyperdoctrine Semantics

Definition 4 *if $\langle \tau_1 \mid X_1 \rangle$ and $\langle \tau_2 \mid X_2 \rangle$ are composable, then*

$$\langle \tau_2 \mid X_2 \rangle \circ \langle \tau_1 \mid X_1 \rangle := \langle \tau_2(\overline{X_2}/\tau_1) \mid X_1 \rangle$$

That is to say, replace the variables X_2 (taken in order of subscripts) occurring in τ_2 with the symbols constituting τ_1, and view the result as a term in the context X_1.

There's a bit of a problem here, however: if we associate each term with a unique arrow, the composition above doesn't give us a category. Recall that to be a category, each object must have a unique identity arrow. The natural candidate for the identity arrow $T_2 \longrightarrow T_2$, for example, is something of the form $\langle x_1 x_2 \mid x_1, x_2 \rangle$. But it's equally natural to consider $\langle x_3 x_4 \mid x_3, x_4 \rangle$. More generally, the only plausible candidates for identity arrows at T_n are terms of the form $\langle \overline{X} \mid X \rangle$. All such arrows are in fact *left* identities. That is, we have for example that

$$\langle x_1 x_2 \mid x_1, x_2 \rangle \circ \langle a x_3 \mid x_3 \rangle = \langle a x_3 \mid x_3 \rangle$$

But the fact that *all* of these arrows are left identities immediately entails that none of them are right identities. Again this is easy to see by examining the following simple example:

$$\langle x_3 x_4 \mid x_3, x_4 \rangle \circ \langle x_1 x_2 \mid x_1, x_2 \rangle = \langle x_1 x_2 \mid x_1, x_2 \rangle$$

Luckily, this example suggests how to correct the problem. Given the definition of composition, it's clear that applying any of the candidate identity arrows on the right of a term is exactly the same thing as applying a change of variables. Thus, for example, $\langle \tau \mid X \rangle \circ \langle \overline{Y} \mid Y \rangle = \langle \tau(\overline{X}/\overline{Y}) \mid Y \rangle$. So, rather than taking arrows to be terms *simpliciter*, we take arrows to be equivalence classes of terms with equivalence being given by monotone change of variables. More explicitly, we say that $\langle \tau \mid X \rangle$ and $\langle \sigma \mid Y \rangle$ are equivalent when $\langle \sigma \mid Y \rangle = \langle \tau(\overline{X}/\overline{Y}) \mid Y \rangle$.

Lemma 3 *Composition as defined in Definition 4 is well-defined on equivalence classes of terms; furthermore, the types, maps, and identities above, taken with this composition, constitute a category.*

Lemma 4 *If $\langle \tau \mid Y \rangle : T_m \longrightarrow T_n$, then for some σ, $\langle \tau \mid Y \rangle = \langle \sigma \mid x_1, \ldots, x_n \rangle$.*

The set $\{x_1, \ldots, x_n\}$ will play a large role in the remainder. Accordingly, we abbreviate $\{x_1, \ldots, x_n\}$ as X_1^n.

Definition 5 (The Base Category) $\mathcal{B}$ is the category that has the types T_n for objects and that has equivalence classes of terms as arrows.

Definition 6 For each n,

1. $\mathcal{L}(T_n)$ is the set of formulas $\phi \mid X$ with $card(X) = n$. We say that such formulas have type T_n.

2. If $\phi \mid X$ and $\psi \mid Y$ are both in $\mathcal{L}(T_n)$, then we say $\phi \mid X$ and $\psi \mid Y$ are equivalent in $\mathbf{K}$ (written: $\phi \mid X \cong_{\mathbf{K}} \psi \mid Y$) when both $\phi \mid X \vdash_{\mathbf{K}} \psi \mid Y$ and $\psi \mid Y \vdash_{\mathbf{K}} \phi \mid X$.

3. For $\phi \mid X \in \mathcal{L}(T_n)$, we write $[\phi \mid X]_{\mathbf{K}}$ for the $\cong_{\mathbf{K}}$-equivalence class of $\phi \mid X$.

4. We write $\mathbf{S}(T_n)$ for the poset with underlying set containing the classes $[\phi \mid X]$ with $card(X) = n$ and with $[\phi \mid X] \leq^n_{\mathbf{K}} [\psi \mid Y]$ iff $\phi \mid X \vdash_{\mathbf{K}} \psi \mid Y$.

5. For each n, we define the unary operation $'^n$ on $\mathbf{S}(T_n)$ by setting $[\phi \mid X]'^n = [\neg \phi \mid X]$.

We leave it to the reader to check that $\cong_{\mathbf{K}}$ is in fact an equivalence relation. When they can be inferred from context (and they essentially always can) we omit most of the superscripts and subscripts.

Lemma 5 $'$ and $\leq$ are well-defined

Lemma 6 Every member of $\mathbf{S}(T_n)$ can be written in the form $[\phi \mid x_1 \ldots x_n]$.

Lemma 7 For all n, the structure $\langle \mathbf{S}(T_n), \leq, ' \rangle$ is a Boolean algebra.

Lemma 8 $[\phi \mid X] = \top_{card(X)}$ iff $\phi \mid X \in \mathbf{K}$, where $\top_n$ is the supremum of the Boolean algebra $\langle \mathbf{S}(T_n), \leq, ' \rangle$.

2 Limning the remaining structure

We've now seen that $\mathcal{B}$'s objects naturally correspond to certain algebras of formulas. As you might expect, it turns out that $\mathcal{B}$'s arrows naturally correspond to homomorphisms of these algebras. To get an intuition for what this is going to look like, consider a formula like $Rx_1x_2 \mid x_1, x_2, x_3$. Our typing conventions demand that we view this formula as incomplete

Hyperdoctrine Semantics

along dimensions x_1, x_2 and x_3. Picture these as three 'slots' into which name-shaped things can be placed. Now notice that if $\tau \mid X$ is a term and $\text{len}(\tau) = 3$, then the untyped part of the term, τ, just is a sequence of three name-shaped things—which is exactly what $Rx_1x_2 \mid x_1, x_2, x_3$ was looking for! Given such a term, we should be able to apply it to $Rx_1x_2 \mid x_1, x_2, x_3$ to get another formula by replacing each occurrence of x_1 by τ_1, x_2 by τ_2, and so on.

To capture all of this, we might try defining a map in the following way:

$$\text{Given} \quad \langle \tau \mid X \rangle : T_n \longrightarrow T_m,$$
$$\text{and} \quad \phi \mid Y \in \mathcal{L}(T_m),$$
$$\text{let} \quad \mathbf{S}\langle \tau \mid X \rangle : \phi \mid Y \longmapsto \phi(\overline{Y}/\tau) \mid X \in \mathcal{L}(T_n).$$

This isn't quite right, but it's close. Before pointing out the problems, we pause to note one important detail that this proposal *does* get right: $\mathbf{S}$, so-defined, is *contravariant* in the sense that it turns arrows $T_n \longrightarrow T_m$ into something going in the other direction, from $\mathcal{L}(T_m)$ to $\mathcal{L}(T_n)$.

The problem, which we point out before fixing, is that $\phi(\overline{Y}/\tau)$ may not be a proper substitution instance of ϕ. For example, consider the formula $\forall x_1 R x_1 x_2 \mid x_2 \in \mathcal{L}(T_1)$ and the term $x_1 \mid x_1 : T_1 \longrightarrow T_1$. Following the above construction, $\mathbf{S}(x_1 \mid x_1)(\forall x_1 R x_1 x_2 \mid x_2) = \forall x_1 R x_1 x_2 (x_2/x_1) \mid x_1 = \forall x_1 R x_1 x_1 \mid x_1$. But notice that before the substitution, the second place of the relation R was occupied by a free variable, and after the substitution it is occupied by a variable bound by the quantifier $\forall x_1$.

We solve this by recalling that the arrows of $\mathcal{B}$ are identified with *equivalence classes* of terms rather than with individual terms. Thus, instead of applying $\tau \mid X$ to $\phi \mid Y$, we can instead apply some equivalent term $\sigma \mid Z$ such that $\phi(\overline{Y}/\sigma)$ is a proper substitution instance of ϕ. It turns out there's a nice way to do this:

Definition 7 *Let $\phi \mid Y \in \mathcal{L}(T_n)$ and $\langle \tau \mid X \rangle : T_m \longrightarrow T_n$. Let Z be a set of variables with $\text{card}(X) = \text{card}(Z)$ and such that no variable in Z occurs bound in ϕ. Let $\sigma = \tau(\overline{X}/\overline{Z})$. Then $\mathbf{S}\langle \tau \mid X \rangle (\phi \mid Y) = \phi(\overline{Y}/\sigma) \mid Z$.*

Lemma 9 *If $\langle \tau \mid X \rangle : T_m \longrightarrow T_n$, then $\mathbf{S}\langle \tau \mid X \rangle$ is a Boolean algebra homomorphism $\mathbf{S}(T_n) \longrightarrow \mathbf{S}(T_m)$. That is, $\mathbf{S}\langle \tau \mid X \rangle$ is well-defined on equivalence classes and commutes appropriately with the Boolean operations.*

As a corollary to the Lemmas proved so far, we have the following:

Corollary 1 **S** *is a contravariant functor that maps each object of* $\mathcal{B}$ *to a Boolean algebra and each arrow of* $\mathcal{B}$ *to a Boolean algebra homomorphism.*

There's something a bit funny we need to deal with now. **S** does, in fact, map each object of $\mathcal{B}$ to the category of Boolean algebras and each arrow of $\mathcal{B}$ to a Boolean algebra homomorphism. But it's useful (for reasons that will be made clear below) to **not** think of **S** as a functor from $\mathcal{B}$ to the category of Boolean algebras and Boolean algebra homomorphisms (call this category **Bool**. Instead, we will think of **S** as a functor from $\mathcal{B}$ to the category of Boolean algebras and *order-preserving* functions (call this category **BoolMon**. Either way you look at it, the point to observe here is that the functor **S** arose very naturally from structure imposed on $\mathcal{L}$ by the relation of **K**-provability. We now turn to showing that **K**-provability not only imposes structure *within* $\mathbf{S}(T_n)$, but also *among* the algebras $\mathbf{S}(T_n)$ for different values of n.

Before we can observe this structure, however, we again need to introduce a bit of notation. To begin, note that for each variable y, and n-membered set of variables X with $y \notin X$ there is a term-arrow $\langle \overline{X} \mid X \cup \{y\}\rangle : T_{n+1} \longrightarrow T_n$. A natural (and, conveniently, correct) interpretation of term-arrows of the form $\langle \overline{X} \mid X \cup \{y\}\rangle$ is that they are projection onto all-but-one component. From here it's not hard to see the following:

Lemma 10 *For* $n > 0$ *and* $y \notin X$, *there are exactly* n *equivalence classes of arrows of the form* $\langle \overline{X} \mid X \cup \{y\}\rangle : T_n \longrightarrow T_{n-1}$; *one for each component omitted.*

Something like the common 'hat' notation to signal omissions is useful here. Usually, one writes $x_1 \ldots \widehat{x}_j \ldots x_n$ to indicate the sequence $x_1 \ldots x_n$, but with x_j omitted. We'll abbreviate further, however, and just write $\widehat{j}$ for this sequence. Thus $\widehat{j} \mid X_1^n$ is shorthand for the term $x_1 \ldots \widehat{x}_j \ldots x_n \mid x_1 \ldots x_n$. It follows that each equivalence class of arrows of the form $\langle \overline{X} \mid X \cup \{y\}\rangle$ has a unique representative of the form $\langle \widehat{j} \mid X_1^n\rangle$.

The maps $\mathbf{S}\langle \widehat{j} \mid X_1^n\rangle$ are very well behaved. To see this, first recall that by Lemma 6 every class in $\mathbf{S}(T_{n-1})$ has a representative with the form $\phi \mid X_1^n - x_j$. But now observe that $\mathbf{S}\langle \widehat{j} \mid X_1^n\rangle[\phi \mid X_1^n - x_j] = [\phi \mid X_1^n]$. Thus, $\mathbf{S}\langle \widehat{j} \mid X_1^n\rangle$ is essentially just the natural *inclusion function* $\mathbf{S}(T_{n-1}) \hookrightarrow \mathbf{S}(T_n)$.

There are equally natural functions going in the other direction:

Definition 8 *Let* Y *be a set of* $n > 0$ *variables, and* y_j *be the* jth *member of* $\overline{Y}$. *Then* $\Pi_j^n[\phi \mid Y] = [\forall y_j \phi \mid Y - y_j]$.

Hyperdoctrine Semantics

Lemma 11 Π_j^n *is an order-preserving function* $\mathbf{S}(T_n) \longrightarrow \mathbf{S}(T_{n-1})$

We emphasize that the various Π functions are *not*, in general, Boolean algebras homomorphisms. For example, $(\Pi_1^1[Rx_1 \mid x_1])' = [\neg \forall x_1 R x_1 \mid \emptyset]$ while $\Pi_1^1([Rx_1 \mid x_1]') = [\forall x_1 \neg R x_1 \mid \emptyset]$. This explains the bit of funny business mentioned after Corollary 1—the Π functions 'live' in the category of Boolean algebras and order-preserving functions, so if we want to 'see' these functions, it's best to view $\mathbf{S}$ as having this category as its codomain. In the remainder, we will write **BoolMon** for the category of Boolean algebras and order-preserving functions and **Bool** for the usual category of Boolean algebras and Boolean algebra homomorphisms.

Π_j^n and $\mathbf{S}\langle\widehat{j} \mid X_1^n\rangle$, as noted, point in opposite directions. As it turns out, they are related in a much more surprising way as well:

Lemma 12 Π_j^n *is right adjoint to* $\mathbf{S}\langle\widehat{j} \mid X_1^n\rangle$.

Proof. Without loss of generality, let $[\phi \mid X_1^n - x_j] \in \mathbf{S}(T_n)$ and let $[\psi \mid X_1^n] \in \mathbf{S}(T_n)$. Our goal is to show that

$$\mathbf{S}\langle\widehat{j} \mid X_1^n\rangle[\phi \mid X_1^n - x_j] \leq [\psi \mid X_1^n] \quad \Leftrightarrow \quad [\phi \mid X_1^n - x_j] \leq \Pi_j^n[\psi \mid X_1^n]$$

It suffices to show that whenever the statement is well formed, we have

$$\phi \mid X_1^n \vdash_{\mathbf{K}} \psi \mid X_1^n \quad \Leftrightarrow \quad \phi \mid X_1^n - x_j \vdash_{\mathbf{K}} \forall x_j \psi \mid X_1^n - x_j$$

For $\Rightarrow$, assume $\phi \mid X_1^n \vdash_{\mathbf{K}} \psi \mid X_1^n$. Then by Lemma 1, $\phi \to \psi \mid X_1^n \in \mathbf{K}$. Since $\phi \mid X_1^n - x_j$ is well-formed, x_j does not occur free in ϕ. Thus by classical logic, $\phi \to \forall x_j \psi \mid X_1^n - x_j \in \mathbf{K}$. So clearly $\phi \mid X_1^n - x_j \vdash_{\mathbf{K}} \forall x_j \psi \mid X_1^n - x_j$.

For $\Leftarrow$, assume $\phi \mid X_1^n - x_j \vdash_{\mathbf{K}} \forall x_j \psi \mid X_1^n - x_j$. Then by Lemma 2, $\phi \mid X_1^n \vdash_{\mathbf{K}} \forall x_j \psi \mid X_1^n$. Also, since it's clear that x_j is free for x_j in ψ, $\forall x_j \psi \to \psi \mid X_1^n \in \mathbf{K}$. Thus $\phi \mid X_1^n \vdash \psi \mid X_1^n$. □

Note that since $\mathbf{S}\langle\widehat{j} \mid X_1^n\rangle$ is essentially an inclusion of $\mathbf{S}(T_{n-1})$ into $\mathbf{S}(T_n)$, Lemma 12 tells us, in a slogan, that universal quantifications are right adjoint to inclusions. It's worth noting that this result does *not* hold if we think of $\mathbf{S}\langle\widehat{j} \mid X_1^n\rangle$ as a functor whose domain is **Bool**, for the simple reason that Π_j^n is not a functor whose *codomain* is **Bool**.

As you might expect, there are dual results for existential quantification:

Definition 9 *Let Y be a set of $n > 0$ variables, and y_j be the jth member of $\overline{Y}$. Then $\Sigma_j^n[\phi \mid Y] = [\exists y_j \phi \mid Y - y_j]$.*

The proof of the next two lemmas are nice exercises that we encourage the reader to pursue.

Lemma 13 Σ_j^n is left adjoint to $\mathbf{S}\langle\widehat{j} \mid X_1^n\rangle$.

Lemma 14 *Without loss of generality, let* $[\psi \mid X_1^n] \in \mathbf{S}(T_n)$ *and let* $[\phi \mid X_1^n - x_j] \in \mathbf{S}(T_{n-1})$. *Then*

$$\Pi_j^n(\mathbf{S}\langle\widehat{j} \mid X_1^n\rangle[\phi \mid X_1^n - x_j] \sqcup [\psi \mid X_1^n]) \leq [\phi \mid X_1^n - x_j] \sqcup \Pi_j^n[\psi \mid X_1^n]$$

The next relationship we mention is a bit more subtle, so we'll take a minute to flesh it out. To begin, consider the following diagram showing two different roads from the formula $Rx_1x_2x_3 \mid x_1, x_2, x_3$ to the formula $\forall x_2 Rax_2 b \mid \emptyset$.

$$\begin{array}{ccc}
Rx_1x_2x_3 \mid x_1, x_2, x_3 & \xrightarrow{\Pi_2^3} & \forall x_2 Rx_1x_2x_3 \mid x_1, x_3 \\
{\scriptstyle \mathbf{S}\langle ax_2b|x_2\rangle}\downarrow & & \downarrow{\scriptstyle \mathbf{S}\langle ab|\emptyset\rangle} \\
Rax_2b \mid x_2 & \xrightarrow{\Pi_1^1} & \forall x_2 Rax_2 b \mid \emptyset
\end{array}$$

It's clear enough, in fact, that the two roads are the same not only at the level of formulas, but also at the level of equivalence classes. That is, the following also commutes:

$$\begin{array}{ccc}
[Rx_1x_2x_3 \mid x_1, x_2, x_3] & \xrightarrow{\Pi_2^3} & [\forall x_2 Rx_1x_2x_3 \mid x_1, x_3] \\
{\scriptstyle \mathbf{S}\langle ax_1b|x_1\rangle}\downarrow & & \downarrow{\scriptstyle \mathbf{S}\langle ab|\emptyset\rangle} \\
[Rax_1 b \mid x_1] & \xrightarrow{\Pi_1^1} & [\forall x_2 Rax_2 b \mid \emptyset]
\end{array}$$

There's nothing special about the particular formulas here, we have as a more general fact that

Lemma 15 *The following diagram commutes:*

$$\begin{array}{ccc}
\mathbf{S}(T_3) & \xrightarrow{\Pi_2^3} & \mathbf{S}(T_2) \\
{\scriptstyle \mathbf{S}\langle ax_1b|x_1\rangle}\downarrow & & \downarrow{\scriptstyle \mathbf{S}\langle ab|\emptyset\rangle} \\
\mathbf{S}(T_1) & \xrightarrow{\Pi_1^1} & \mathbf{S}(T_0)
\end{array}$$

Proof. Without loss of generality, let $[\phi \mid x_1, x_2, x_3] \in \mathbf{S}(T_3)$. Following the 'top' path is easy and takes us to $[\forall x_2 \phi(x_1/a, x_3/b) \mid \emptyset]$. Following the 'bottom' path we first get to $[\phi(x_1/a, x_2/x_1, x_3/b) \mid x_1]$, and from there to $[\forall x_1 \phi(x_1/a, x_2/x_1, x_3/b) \mid \emptyset]$. But clearly this class intersects to $[\forall x_2 \phi(x_1/a, x_3/b) \mid \emptyset]$. So since equivalence classes are disjoint, these are in fact the same class. □

The general result being exemplified here is the following:

Lemma 16 *Without loss of generality, let $\langle \tau \mid X_1^n \rangle : T^n \longrightarrow T^m$ be a term. Then the following commutes:*

$$\begin{array}{ccc} \mathbf{S}(T_{m+1}) & \xrightarrow{\Pi_j^{m+1}} & \mathbf{S}(T_m) \\ {\scriptstyle \mathbf{S}\langle \overline{\tau_{<j}} x_{n+1} \overline{\tau_{\geq j}} \mid X_1^{n+1} \rangle} \Big\downarrow & & \Big\downarrow {\scriptstyle \mathbf{S}\langle \tau \mid X_1^n \rangle} \\ \mathbf{S}(T_{n+1}) & \xrightarrow[\Pi_{n+1}^{n+1}]{} & \mathbf{S}(T_n) \end{array}$$

Proof. Without loss of generality, let $[\phi \mid X_1^{m+1}] \in \mathbf{S}(T_{m+1})$. Via the top path, this class gets sent to $[\forall x_j \phi(\overline{X_1^{m+1} - x_j}/\tau) \mid X_1^n]$. Via the bottom path, it gets sent to $[\forall x_{n+1} \phi(\overline{X_1^{j-1}}/\tau_{<j}, x_j/x_{n+1}, \overline{X_{j+1}^{m+1}}/\tau_{>j}) \mid X_1^n]$. To see these are the same class, observe that the first can be rewritten as $[\forall x_j \phi(\overline{X_1^{j-1}}/\tau_{<j}, x_j/x_j, \overline{X_{j+1}^{m+1}}/\tau_{>j}) \mid X_1^n]$. With by-now-familiar tricks, we then see that if y occurs in neither representative formula, then both prove $\forall y \phi(\overline{X_1^{j-1}}/\tau_{<j}, x_j/y, \overline{X_{j+1}^{m+1}}/\tau_{>j}) \mid X_1^n$. Equally clearly, and by the same tricks, this formula proves both representatives. Thus the classes intersect, so are identical. □

3 Hyperdoctrines

That was a lot of information. Here are what we take to be the important bits:

Corollary 1 $\mathbf{S}$ is a contravariant functor from $\mathcal{B}$ to **BoolMon** whose image is in **Bool**.

Lemma 12 *Each arrow of the form $\mathbf{S}\langle \widehat{j} \mid X_1^{n+1} \rangle$ has a right adjoint Π_j^{n+1}.*

Lemma 14 *Whenever all of it makes sense, we get that*

$$\Pi_j^n(\mathbf{S}\langle \widehat{j} \mid X_1^n \rangle [\phi \mid X_1^n - x_j] \sqcup [\psi \mid X_1^n]) \leq [\phi \mid X_1^n - x_j] \sqcup \Pi_j^n [\psi \mid X_1^n]$$

Lemma 16 *Whenever all of it makes sense, the following diagram commutes:*

$$\begin{array}{ccc} \mathbf{S}(T_{m+1}) & \xrightarrow{\Pi_j^{m+1}} & \mathbf{S}(T_m) \\ {\scriptstyle \mathbf{S}\langle \overline{\tau_{<j}} x_{n+1} \overline{\tau_{\geq j}} | X_1^{n+1} \rangle} \downarrow & & \downarrow {\scriptstyle \mathbf{S}\langle \tau | X_1^n \rangle} \\ \mathbf{S}(T_{n+1}) & \xrightarrow[\Pi_{n+1}^{n+1}]{} & \mathbf{S}(T_n) \end{array}$$

We define a Boolean hyperdoctrine to be a functor that has 'the same structure' as the functor $\mathbf{S}$:

Definition 10 *A Boolean hyperdoctrine is a contravariant functor $H : \mathcal{B} \longrightarrow \mathbf{BoolMon}$ such that*[4]

BH1 *The image of H is in* **Bool**

BH2 *Each arrow of the form $H \langle \widehat{j} \mid X_1^{n+1} \rangle$ has a right adjoint Π_j^{n+1}.*

BH3 *Whenever all of it makes sense, we get that*

$$\Pi_j^n(H\langle \widehat{j} \mid X_1^n \rangle [\phi \mid X_1^n - x_j] \sqcup [\psi \mid X_1^n]) \leq [\phi \mid X_1^n - x_j] \sqcup \Pi_j^n[\psi \mid X_1^n]$$

BH4 *Whenever all of it makes sense, the following diagram commutes:*

$$\begin{array}{ccc} H(T_{m+1}) & \xrightarrow{\Pi_j^{m+1}} & H(T_m) \\ {\scriptstyle H\langle \overline{\tau_{<j}} x_{n+1} \overline{\tau_{\geq j}} | X_1^{n+1} \rangle} \downarrow & & \downarrow {\scriptstyle H\langle \tau | X_1^n \rangle} \\ H(T_{n+1}) & \xrightarrow[\Pi_{n+1}^{n+1}]{} & H(T_n) \end{array}$$

When it matters, we will distinguish the elements of and operations on the various algebras $H(T_n)$ by subscripting them. E.g. if necessary we will write $\leq_n$ for the partial order in $H(T_n)$ or $\top_n$ for its top element. Two key differences between arbitrary Booleans hyperdoctrine and the *syntactic hyperdoctrine* $\mathbf{S}$ are worth noting explicitly:

---

[4]There is a more general version of hyperdoctrines that take as their domain *any* category with enough structure to interpret the types of the language in question. This added complication adds little of importance in the case at hand, so is ignored.

Hyperdoctrine Semantics

- In **S**, the algebras $\mathbf{S}(T_n)$ are always algebras of formulas. In an arbitrary Boolean hyperdoctrine H, $H(T_n)$ can be any Boolean algebra whatsoever.

- In **S**, the homomorphisms $\mathbf{S}\langle \tau \mid X \rangle$ always arise via substitution. In an arbitrary Boolean hyperdoctrine H, $H\langle \tau \mid X \rangle$ can be any Boolean algebra homomorphism whatsoever.

Definition 11 *If H is a Boolean hyperdoctrine, then an interpretation of $\mathcal{L}$ in H is a function $[\![-]\!]$ that assigns a member $[\![R]\!] \in H(T_n)$ to each n-ary predicate R. An interpretation induces an assignment of a semantic value $[\![\phi \mid X]\!]$ to each formula in the following way:*

- *If R is m-ary and $\tau \mid X : T_n \longrightarrow T_m$, then $[\![R\tau \mid X]\!] = H[\![\tau \mid X]\!][\![R]\!]$.*
- $[\![\neg \phi \mid X]\!] = [\![\phi \mid X]\!]'$
- $[\![\phi \wedge \psi \mid X]\!] = [\![\phi \mid X]\!] \sqcap [\![\psi \mid X]\!]$
- $[\![\phi \vee \psi \mid X]\!] = [\![\phi \mid X]\!] \sqcup [\![\psi \mid X]\!]$
- $[\![\phi \to \psi \mid X]\!] = [\![\phi \mid X]\!]' \sqcup [\![\psi \mid X]\!]$
- *If $\overline{Y} = y_1 y_2 \ldots y_n$, then $[\![\forall y_j \psi \mid Y]\!] = \Pi_j^n [\![\psi \mid Y - y_j]\!]$.*

Definition 12 *The identity interpretation for the syntactic hyperdoctrine— written $[\![-]\!]_{\mathrm{id}}$— is the assignment $[\![R]\!]_{\mathrm{id}} = [R\overline{X_n^1} \mid X_n^1]$.*

Lemma 17 $[\![\phi \mid X]\!]_{\mathrm{id}} = [\phi \mid X]$ *for all formulas $\phi \mid X$.*

Lemma 18 *For any Boolean hyperdoctrine H and interpretation $[\![-]\!]$,*

$$[\![\langle \tau \mid X \rangle(\phi \mid Y)]\!] = H\langle \tau \mid X \rangle[\![\phi \mid Y]\!]$$

Definition 13

- *We say that $\phi \mid X$ is **K**-valid in H relative to the interpretation $[\![-]\!]$ when $[\![\phi \mid X]\!] = \top_{\mathrm{card}(X)}$.*

- *We say $\phi \mid X$ is **K**-valid in H when $\phi \mid X$ is **K**-valid in H relative to every interpretation.*

- *We say $\phi \mid X$ is **K**-valid when $\phi \mid X$ is **K**-valid in H for every H.*

Theorem 1 *If $\phi \mid X$ is **K**-valid, then $\phi \mid X \in \mathbf{K}$.*

Proof. We prove the contrapositive. If $\phi \mid X \notin \mathbf{K}$, then by Lemma 8, $[\![\phi \mid X]\!] \neq \top_{\mathrm{card}(X)}$. So by Lemma 17, $\phi \mid X$ is not **K**-valid in the syntactic hyperdoctrine equipped with the identity interpretation. Thus $\phi \mid X$ is not **K**-valid in the syntactic hyperdoctrine. So $\phi \mid X$ is not **K**-valid. □

Corollary 2 *If $[\![\phi \mid X]\!] \leq [\![\psi \mid X]\!]$ in every interpreted Boolean hyperdoctrine, then $\phi \mid X \vdash_{\mathbf{K}} \psi \mid X$.*

It follows from Theorem 1 and Corollary 2 that the structural features we've identified in B1-B4 are satisfied by enough structures to falsify every nontheorem and to counterexample every nonentailment. But the syntactic hyperdoctrine has many features that we *didn't* include in B1-B4. As an example, not only is it true that the various algebras $\mathbf{S}(T_n)$ are Boolean algebras, it also happens to be the case that they are all countably infinite Boolean algebras generated by a countable infinity of atoms. But B1 doesn't require *any* of the $H(T_n)$'s to have either of these features. So B1-B4 allow us to interpret $\mathcal{L}$ in structures that look quite different from the prototypical example of a Boolean hyperdoctrine, **S**.

Of course, admitting Boolean hyperdoctrines that are quite different from **S** runs the risk of admitting Boolean hyperdoctrines in which we can interpret some *theorems* as false and/or counterexample some entailments. That this *doesn't* happen is, we think, somewhat surprising.

Theorem 2 *If $\phi \mid X \in \mathbf{K}$, then $\phi \mid X$ is **K**-valid.*

The proof, which we omit, is a straightforward induction on the length of the proof witnessing that $\phi \mid X \in \mathbf{K}$.

Corollary 3 *If $\phi \mid X \vdash_{\mathbf{K}} \psi \mid X$, then in every interpreted Boolean hyperdoctrine $[\![\phi \mid X]\!] \leq [\![\psi \mid X]\!]$.*

4 Metaphysical interpretation

We claimed at the outset that the story we were telling here would be of interest to metaphysicians. In this section we finally make good on that promise. The novelty of the presentation above is this: we have presented a respectable semantics for quantified first-order logic without any appeal to things that are being quantified over. We have, if you like, Being ($\exists$, that is) without beings.

Ordinary model theory is emphatically not like this. In an ordinary model (of the signature Σ, say) one has first of all, a domain of individuals. One then has the interpretations of the various symbols of Σ: interpretations of predicates are sets of individuals, interpretations of constants are individuals, and so on. Thus, the whole model-theoretic edifice *grounds out*, in some sense, at the level of individuals.

The grounding going on here, whatever it might be, is a fairly robust matter. To begin, there is the obvious dependence of sets on their members. But even putting that to the side, the comparison of structures ultimately comes down to the sets of individuals in their domains; maps of Σ-structures are simply maps between the underlying domains of those structures that happen to have certain further properties. The grounding of the model-theoretic world on the world of individuals and particulars further reveals itself on even a casual examination of many of the classical results of the subject. As often as not, said results are either statements about possible cardinalities for structures, or statements about how many structures there are (up to isomorphism) of a certain cardinality. Making generalizations about the psychology of workers in a scientific field is a risky business, but it seems fair enough to say that the models are fundamentally understood to be decorated sets (like groups, fields, and other objects in concrete categories), and that their underlying sets and the individuals that inhabit them are fundamental to the subject.

This incursion of set theoretic concepts into model theory, and from there into metaphysics, where sets are smuggled in as indispensable for semantics, has deeply colored contemporary analytic philosophy, both subtly and overtly. We can give some examples of both kinds of coloring.

Among the overtly colored subjects, we have, for example, the family of problems related to absolute generality. Parsons describes one of these problems in the following way:

> The universe of [the] metaphysician's purview surely includes everything, with no restriction tacit or otherwise. Logic might seem at first sight to envision only restricted generalization. We interpret the language of quantified logic with respect to a domain or 'universe of discourse'... Typically the domain is a set, and set theory tells us how, given a set, to describe a set containing elements not in the first set. In a sense, the received way of interpreting quantificational logic takes all quantifiers to be restricted. (Parsons, 2006, p. 203)

This is to put everything a little plainly, and in the literature one finds other approaches to quantification—holding on to the standard semantic machinery but replacing sets with classes, properties, pluralities, or some other kind of collection, or finding some way to think of many domains as being stitched together into one, perhaps via the semantics of modal logic—but the basic model-theoretic flavoring is clear. If one does without the idea of a domain (as we do) then it is not clear that the problem of absolute generality is even expressible, let alone a problem.

One finds overt model theoretic flavoring in certain versions of the bad company objection, familiar to Neologicists (Boolos, 1987). Here, the problem is roughly as follows. Certain axioms that Neologicists would like to have can only be true in models with domains of certain cardinalities. Sometimes, the cardinalities allowed by one axiom do not overlap with the cardinalities allowed by another. At this point, it's generally taken to be clear that the axioms are incompatible in some metaphysically deep sense, even though it's quite open (since we are dealing with second-order model theoretic semantics, for which there is no completeness theorem) that the axioms are perfectly consistent with one another in spite of not being jointly satisfiable.

A third example might be found in the debate over Putnam's model-theoretic argument for anti-realism (Putnam, 1980), and more generally in the discussion surrounding Skolem's paradox. Putnam argues roughly that, since by standard model theoretic results, there are many first-order models (of varying cardinalities) in which any set of platitudes or observations we might put forward would be satisfied, much of our mathematical language cannot have a fixed "intended interpretation", to the point where statements independent of the Zermelo-Frankel axioms cannot have truth values. Putnam proposes a "non-realist semantics" to get around this. The point at hand thought is that Putnam simply *assumes* that realism entails some form of broadly model-theoretic semantics for natural language.

As for less overt colorings, one could multiply examples endlessly. Lewis' metaphysics has a broadly model-theoretic flavor from the identification of properties with sets of (possible) individuals in On the Plurality of Worlds up through the fairly explicit picture of language in General Semantics.[5] And within metaphysics, Lewis casts a long shadow. More generally, the idea that the ground floor of metaphysics should somehow be a set of discrete individuals of some kind (whether mereological fusions, simple substances, space-time points, events, tropes, or some other kind of thing), over which

---

[5] See (Lewis, 1986) and (Lewis, 1970).

Hyperdoctrine Semantics

our best theory quantifies, is ubiquitous. And it's this picture to which we are offering an alternative.

What is the alternative? It's a world in which the ground floor is not a bunch of things over which we quantify, but a bunch of propositions, upon which the quantifiers act, transforming them from type to type. The world, at least as far as logic is concerned, is a totality of facts, not of things. Or, more precisely, the world is a totality of propositional functions organized into families by type. All that remains is determining how these families hang together. So—and this is the fun part—objects are in an important sense *secondary* features of the world, emergent from the underlying propositional structure. More to the point, objects on the hyperdoctrinal perspective are homomorphisms of algebras of propositional functions rather than members of a domain of quantification. Thus it is the *algebras* and not the *objects* that are taken as primitives—the whole edifice, that is, grounds out at the level of the algebra of propositional functions, not at the level of individuals. Objects, of course, still play an important role in our thinking about how the families of propositional functions hang together. But in the same way you would want to say that an isomorphism (a function) between two structures exists because of the way the structures are, rather than explaining the way the structures are by appealing to the existence of a certain function, one can say that objects (regarded as a certain type of homomorphism) depend on the algebraic structure of propositions, rather than the other way around.

One might object here that, even if hyperdoctrinal semantics lets us do first-order logic without a commitment to objects, it still commits us to a whole zoo of categorial machinery. The exposition of the theory commits us to functors, to adjoint pairs, to categories themselves...This objection however, misunderstands the nature of the machinery. It's the following kind of mistake: Imagine a nominalist who, upon discovering that his pet bird was in fact, an African swallow, bemoaned his new ontological commitment to African swallows. When we call a bird a swallow, we just give it a name that conceptualizes it as part of an orderly scheme for classifying organisms— we don't postulate a new thing. Analogously, functors, adjoint pairs, and categories are just ways of organizing and conceptualizing familiar structure, not new categories of beings that we here postulate. Our basic commitments are only to the consequence relation, to the propositional functions it orders, and to some basic operations on propositional functions, like negation and quantification.

Furthermore, hyperdoctrinal semantics is neutral on the nature of consequence, and on how to account for the algebraic relations of propositional functions. So, there is room here for a variety of different metaphysical pictures; perhaps one could return to the idea of objects inhabiting a domain of quantification if this seems the best way to explain what it means for one propositional function to entail another. The point is that this layer of metaphysical structure is not at all required for a precise metatheory of first-order logic. Instead, it is up to the metaphysician to motivate it, or reject it, on grounds internal to their practice rather than by appeal to some alleged logical necessity.

And there are more degrees of freedom here than just the recovery or abandonment of standard model-theoretic semantics, because there's no particular reason to restrict to the usual metaphysical data. Categories are agnostic about the structure of the objects they're made up of. This agnosticism lifts, in the case at hand, to an agnosticism about how ontology ought to be done. Thus, if agnosticism appeals to you, you ought to find hyperdoctrinal semantics a welcoming space. If you are agnostic about agnosticism, you're still likely to find hyperdoctrinal semantics useful, as a way of disentangling your gnostic ruminations from your logical commitments, and freeing up more space for you to explore.

References

Boolos, G. (1987). The consistency of Frege's Foundations of Arithmetic. In J. J. Thompson (Ed.), *On Being and Saying: Essays in Honor of Richard Cartwright* (pp. 3–20). Cambridge, MA: MIT Press.

Lewis, D. K. (1970). General semantics. *Synthese*, *22*(1-2), 18–67.

Lewis, D. K. (1986). *On the Plurality of Worlds*. Oxford: Blackwell.

Parsons, C. (2006). The problem of absolute universality. In A. Rayo & G. Uzquiano (Eds.), *Absolute Generality*. Oxford: Oxford University Press.

Putnam, H. (1980). Models and reality. *Journal of Symbolic Logic*, *45*(3), 464–482.

Shay Logan
Kansas State University, Department of Philosophy
United States
E-mail: salogan@ksu.edu

Graham Leach-Krouse

Hyperdoctrine Semantics

Kansas State University, Department of Philosophy
United States
E-mail: gleachkr@ksu.edu

Truthmaker Semantics for Infectious Logics

THOMAS RANDRIAMAHAZAKA[1]

Abstract: This paper introduces and motivates a new truthmaker semantics for disjunction. It gives rise to sound and complete inexact truthmaker semantics for Weak Kleene Logic and Paraconsistent Weak Kleene Logic. The semantics is then generalised to capture all truth-functional infectious logics. We then discuss the philosophical upshot of the formal systems presented.

Keywords: truthmaker semantics, weak Kleene, infectious logics, nonsense logics

1 Introduction

Truthmaker semantics is a recent development of truth-conditional formal semantics (e.g., Fine, 2017a, 2017b; Leitgeb, 2019). One can find two main motivations for its study. First, truthmaker semantics can provide a fine-grained, hyperintensional representation of propositional contents. Such a representation can then be used to devise more realistic accounts of propositional attitudes (e.g., Barwise & Perry, 1981; Hawke & Özgün, in press; Korbmacher, Anglberger, & Faroldi, 2016). Truthmaker semantics constitutes then a generalisation of and an improvement upon the standard possible world account of propositional content. Second, truthmaker semantics can be used to provide sound and complete semantics for non-classical logics (e.g., Fine, 2016; Jago, 2020; Van Fraassen, 1969). Here, the level of granularity offered by truthmaker semantics can be used to capture the subtle differences between propositions that classical logic identifies. These two motivations are heavily interconnected. The fact that the truthmaker semantics used to provide an account of this particular propositional attitude happens to be sound and complete with respect to that particular logic offers a valuable insight into the internal grammar of this propositional attitude. Similarly, the

---

[1] I would like to thank Franz Berto for his valuable comments.

fact that a certain account of propositional content can be modelled by the truthmaker semantics sound and complete for a particular logic provides a deep semantic understanding of that logic. We hope to achieve the later in this paper.

There is a great variety of different truthmaker semantics in the literature. One can however observe that they tend to agree, *mutatis mutandis*, on their treatment of disjunction. In this paper, we consider a motivated alternative semantic treatment of disjunction. In Section 2, we present and motivate the alternative semantic clause. In Section 3, we present a formal system which integrates the new clause and study the two logics it induces. These two logics happen to be Weak Kleene Logic and Paraconsistent Weak Kleene Logic. These two logics are the paradigmatic examples of infectious logics. In Section 4, we show how to generalise the system presented in Section 3 to provide sound and complete truthmaker semantics for all truth-functional infectious logics. In Section 5, we discuss how the formal systems developed in this paper can help to cast a new understanding on Weak Kleene Logic and Paraconsistent Kleene Logic in particular but also on infectious logics more generally. An appendix will contain all the proofs.

2 Towards a new semantics for disjunction

A truthmaker for a statement P is a particular way for P to be true (Yablo, 2014). Similarly, a falsitymaker for P is a particular way for P to be false. The idea of a recursive truthmaker semantics for a n-ary connective $\dagger$ consists then in that, once one knows the particular ways in which the statements $P_1, ..., P_n$ can be true or false, one is in position to determine – via a precise algorithm – the particular ways in which the complex statement $\dagger(P_1, ..., P_n)$ can be true or false. This algorithm is to be encapsulated in the truthmaking and falsitymaking clauses for $\dagger$.

For instance, a conjunction is true if both its conjuncts are true. Therefore, a particular way for the statement $P \wedge Q$ to be true must be a combination of a particular way for P to be true with a particular way for Q to be true. There are, however, two readings of this talk of combination. According to the first one, which we dub simultaneity reading, a combination of a particular way for P to be true with a particular way for Q to be true is a situation which is simultaneously *this* particular way for P to be true and *that* particular way for Q to be true. By contrast, under the second reading that we call mereological reading, a combination of a particular way for P to be true with a particular way for Q to be true is the mereological sum of a situation which is *this* particular way for P to be true with a situation which is *that* particular way for Q to be true. Clearly, the simultaneity reading is stronger than the mereological one.

The distinction between the simultaneity reading and the mereological reading often coincides with the distinction between two interpretations of the truthmaking/falsitymaking relations, that is to say two ways of understanding what it means to be a particular way for a statement to be being true/false. The word *to be* in "to be a particular way for P to be true" can be understood as expressing an identity or a predication. If it is an identity, then a truthmaker for P will be relevant as a whole to the truth of P. If it is a predication, then a truthmaker for P might only be partially relevant to the truth of P. Indeed, a property exemplified by an object generally only partially characterises it, whereas the relation of identity concerns the object as whole. These two types of truthmaking are called exact and inexact truthmaking, respectively (Deigan, 2020; Fine, 2017a; Leitgeb, 2021).[2] To disambiguate, we will say that an exact truthmaker for P *constitutes* a particular way for P to be true whereas an inexact truthmaker for P *exemplifies* a particular way for P to be true. Formally, the main difference between the two is that any mereological extension of an inexact truthmaker for P will itself be an inexact truthmaker for P whereas this is not guaranteed in the exact case. The reason is that, in the inexact case, a situation is said to exemplify a particular way for P to be true in virtue of what is going on inside that situation, and therefore every mereological extension of that situation will exemplifies the very same particular way for P to be true. On the other hand, nothing ensures that a situation and its mereological extensions will constitute the same way for P to be true.

Notice that the simultaneity and mereological readings are equivalent under the inexact interpretation. It suffices to notice that the combination of a particular way for P to be true with a particular way for Q to be true (under the mereological reading) is a mereological extension of both a particular way for P to be true and a particular way for Q to be true. Because it leads to simpler formal systems, it is understandable that inexact

---

[2] As noted by an anonymous reviewer, truthmaker semantics has recently been mostly associated with Fine's work and therefore with exact truthmaking. For this reason, some might disagree with the application of the label "truthmaker semantics" to an inexact semantics. However, Fine himself makes this distinction and the debate concerning the philosophical priority between exact and inexact truthmaking is cashed out in precisely these terms, not between "truthmaker semantics" and something else.

truthmaking semantics tend to choose the simultaneity reading. Consequently, the usual inexact truthmaking clause for conjunction is the following, where $\Vdash^+$ represents the relation of truthmaking:

$$s \Vdash^+ \varphi \wedge \psi \text{ if and only if } s \Vdash^+ \varphi \text{ and } s \Vdash^+ \psi$$

Under the exact reading however, they are clearly distinct. Because there is no guarantee, for instance, that being wholly relevant to the truth of a conjunction entails being wholly relevant to the truth of each of its conjuncts, the mereological reading is more adequate for an exact truthmaker semantics. Therefore, the usual exact truthmaking clause for conjunction is the following, where $\sqcup$ represents mereological fusion:

$$s \Vdash^+ \varphi \wedge \psi \text{ if and only if } s = s_1 \sqcup s_2 \text{ and } s_1 \Vdash^+ \varphi \text{ and } s_2 \Vdash^+ \psi$$

When aiming to determine all truthmakers for a statement P, there are some constraints to one might have in mind. The first one is Exhaustivity: if P is true it must be in virtue of the obtaining of a particular way for P to be true. Exhaustivity demands an obtaining truthmaker for every true proposition. Another, maybe more controversial, condition is Exclusivity: at most one particular way for P to be true obtains. In an exact setting, this means that at most one truthmaker of P obtains. In an inexact setting, it means that if two truthmakers of P obtains then they exemplify the same way for P to be true. The idea behind Exclusivity is that we are talking about *particular* ways of being true and that this particularity prevents overdetermination. In the classical setting of possible world semantics, accepting Exhaustivity and Exclusivity can be seen as functionally identifying the set of ways for P to be true with a partition of the P-worlds.[3] It is not the goal of this paper to argue for Exhaustivity and Exclusivity but, rather, we would like to see what kind of semantics one gets when these two constraints are taken seriously. The spirit is experimental.

Exclusivity clashes with the traditional understanding of disjunction according to which one distinguishes two cases in which $P \vee Q$ is true: the situations in which P is true and the situations in which Q is true. The notion of combination being absent, both the simultaneity and mereological readings give rise to the following clause:

$$s \Vdash^+ \varphi \vee \psi \text{ if and only if } s \Vdash^+ \varphi \text{ or } s \Vdash^+ \psi$$

---

[3]This follows (Yablo, 2014)'s conception of truthmaking.

But suppose that both P and Q are true. Then, surely, $P \vee Q$ is true but it is not clear if we are in the first case or in the second case. Clearly, we are in both and that is the problem, assuming Exclusivity. A similar problem of overdetermination can be found in the grounding literature. More particularly, if one is looking for a notion of unique immediate logical ground, then one can wonder what it is for $P \vee Q$ in the cases in which both P and Q are true. The traditional understanding of disjunction cannot yield the desired uniqueness. The solution proposed by Genco, Poggiolesi, and Rossi (2021) is to distinguish three - and not two - different possibilities for the immediate logical ground of $P \vee Q$: the truth P alone, the truth Q alone and the truths of P and Q together. The truth of P alone is then understood as the truth of P under the condition that Q is false. Similarly, the truth of Q alone is understood as the truth of Q under the condition that P is false. Uniqueness follows. Following that idea, we can distinguish three cases in which $P \vee Q$ is true: the situations in which P is true but Q is false, the situations in which Q is true but P is false and the situations in which P and Q are both true. Accordingly, a particular way for $P \vee Q$ to be true must take one of those three following form:

(a) a combination of a particular way for P to be true with a particular way for Q to be false,

(b) a combination of a particular way for P to be false with a particular way for Q to be true,

(c) a combination of a particular way for P to be true with a particular way for Q to be true.

The simultaneity reading then produces the following clause (where $\Vdash^-$ represents falsitymaking):

$$s \Vdash^+ \varphi \vee \psi \text{ iff } \begin{cases} s \Vdash^+ \varphi \text{ and } s \Vdash^- \psi, \text{ or} \\ s \Vdash^- \varphi \text{ and } s \Vdash^+ \psi, \text{ or} \\ s \Vdash^+ \varphi \text{ and } s \Vdash^+ \psi. \end{cases}$$

The mereological reading, by contrast, produces the following clause:

$$s \Vdash^+ \varphi \vee \psi \text{ iff } s = s_1 \sqcup s_2 \text{ where } \begin{cases} s_1 \Vdash^+ \varphi \text{ and } s_2 \Vdash^- \psi, \text{ or} \\ s_1 \Vdash^- \varphi \text{ and } s_2 \Vdash^+ \psi, \text{ or} \\ s_1 \Vdash^+ \varphi \text{ and } s_2 \Vdash^+ \psi. \end{cases}$$

The rest of this paper is dedicated to inexact truthmaker semantics integrating the simultaneity reading of this new clause for disjunction. Consequently, we leave for future work the study of exact truthmaker semantics integrating the mereological reading of this new clause.

3 Possibilist truthmaker semantics

In an inexact context where every statement is either true or false, like possible world semantics, our new clause is equivalent to the traditional understanding of disjunction. They differ significantly however in contexts where bivalence can fail at the level of truth-conditions. For instance, a partial situation can make a statement P true without deciding the truth or the falsity of another statement Q. Such a situation would make $P \vee Q$ true under the traditional understanding of disjunction but not under the new one. What kind of logic one gets with such a semantics? Let us now answer that question.

The formal language we consider is the classical language $\mathcal{L}_c$ recursively generated from a countably infinite set $Prop$ of propositional letters with conjunction ($\wedge$), disjunction ($\vee$) and negation ($\neg$). We need to have a truthmaking and a falsitymaking clause for each of these connectives. The truthmaking clause for disjunction has obviously already been discussed. The truthmaking clause for conjunction is going to be the traditional inexact clause for conjunction also already mentioned. The falsitymaking clause for disjunction is going to be the dual of the truthmaking clause for conjunction, to ensure De Morgan duality. Similarly, the falsitymaking clause for conjunction is going to be the dual of the truthmaking clause for disjunction. Therefore, our semantics is non-standard not only because of its treatment of disjunction (the truthmaking clause) but also for its treatment of conjunction (the falsitymaking clause). The truthmaking and falsitymaking clauses for negation just switch between truthmaking and falsitymaking.

As mentioned, the states which are to be truthmakers and falsitymakers are going to be partial situations. Otherwise, the new clause for disjunction is equivalent to the traditional one. The simplest case of such states take them to be possible and, therefore, consistent. Think of mereological parts of possible worlds according to modal realism. Obviously, the new treatment of disjunction does not require such a restriction and could be be applied to inconsistent, impossible situations. We leave the study of such systems to future research.

Truthmaker Semantics for Infectious Logics

Let's go formal. A possibilist frame is a partial order $\langle S, \sqsubseteq \rangle$. The elements of S are called states and $\sqsubseteq$ represents the mereological order on S. For all states $s, t \in S$, we say that s and t are compatible if they have a common mereological extension, i.e. if there is some $u \in S$ such that $s \sqsubseteq u$ and $t \sqsubseteq u$. Otherwise, we say that they are incompatible. The intended interpretation of possibilist frames is constituted by possible situations ordered by parthood.

A possibilist model is a tuple $\langle S, \sqsubseteq, v^+, v^- \rangle$ where $\langle S, \sqsubseteq \rangle$ is a possibilist frame and, for $\circ \in \{+, -\}$, we have $v^\circ : Prop \to \mathcal{P}(S)$ such that, for all $p \in Prop$ and $s, t \in S$, (a) if $s \in v^\circ(p)$ and $s \sqsubseteq t$ then $t \in v^\circ(p)$ and (b) if $s \in v^+(p)$ and $t \in v^-(p)$ then s and t are incompatible. The set $v^+(p)$ represents the truthmakers of p and the set $v^-(p)$ represents its falsitymakers. The first condition on v^+ and v^- ensures that truth and falsity are preserved by mereological extension. The second rules out inconsistent states.

We can now define the truthmaking and falsitymaking relations for all formulas of $\mathcal{L}_c$:

- $s \Vdash^+ p$ iff $s \in v^+(p)$,
- $s \Vdash^- p$ iff $s \in v^-(p)$,
- $s \Vdash^+ \varphi \wedge \psi$ iff $s \Vdash^+ \varphi$ and $s \Vdash^+ \psi$,
- $s \Vdash^- \varphi \wedge \psi$ iff $\begin{cases} s \Vdash^- \varphi \text{ and } s \Vdash^+ \psi, \text{ or} \\ s \Vdash^+ \varphi \text{ and } s \Vdash^- \psi, \text{ or} \\ s \Vdash^- \varphi \text{ and } s \Vdash^- \psi, \end{cases}$
- $s \Vdash^+ \varphi \vee \psi$ iff $\begin{cases} s \Vdash^+ \varphi \text{ and } s \Vdash^- \psi, \text{ or} \\ s \Vdash^- \varphi \text{ and } s \Vdash^+ \psi, \text{ or} \\ s \Vdash^+ \varphi \text{ and } s \Vdash^+ \psi, \end{cases}$
- $s \Vdash^- \varphi \vee \psi$ iff $s \Vdash^- \varphi$ and $s \Vdash^- \psi$,
- $s \Vdash^+ \neg \varphi$ iff $s \Vdash^- \varphi$,
- $s \Vdash^- \neg \varphi$ iff $s \Vdash^+ \varphi$.

The first thing to check is that the two conditions imposed on the truthmakers and falsitymakers of propositional letters hold for all formulas of $\mathcal{L}_c$.

Proposition 1 *For all formulas φ and states $s, t \in S$ such that $s \sqsubseteq t$, if $s \Vdash^+ \varphi$ then $t \Vdash^+ \varphi$ and if $s \Vdash^- \varphi$ then $t \Vdash^- \varphi$.*

Proposition 2 *For all formulas φ and states $s, t \in S$, if $s \Vdash^+ \varphi$ and $t \Vdash^- \varphi$ then s and t are incompatible.*

It is possible to define two notions of logical consequence within this semantics, depending on whether one cares about preservation of truth or preservation of non-falsity. A formula φ strictly entails a formula ψ, written $\varphi \models_s \psi$, if all truthmakers of φ are truthmakers of ψ in all possibilist models. By contrast, φ tolerantly entails ψ, written $\varphi \models_t \psi$, if all falsitymakers of ψ are falsitymakers of φ in all possibilist models.[4]

Interestingly, the logic of strict entailment coincides with Weak Kleene Logic (K_3^w). Similarly, the logic of tolerant entailment coincides with Paraconsistent Weak Kleene Logic (PWK) (e.g., Ciuni & Carrara, 2019). These two logics are defined on the Weak Kleene truth tables:

$\wedge$	$\mathfrak{f}$	$\mathfrak{t}$	$\mathfrak{e}$
$\mathfrak{f}$	$\mathfrak{f}$	$\mathfrak{f}$	$\mathfrak{e}$
$\mathfrak{t}$	$\mathfrak{f}$	$\mathfrak{t}$	$\mathfrak{e}$
$\mathfrak{e}$	$\mathfrak{e}$	$\mathfrak{e}$	$\mathfrak{e}$

$\vee$	$\mathfrak{f}$	$\mathfrak{t}$	$\mathfrak{e}$
$\mathfrak{f}$	$\mathfrak{f}$	$\mathfrak{t}$	$\mathfrak{e}$
$\mathfrak{t}$	$\mathfrak{t}$	$\mathfrak{t}$	$\mathfrak{e}$
$\mathfrak{e}$	$\mathfrak{e}$	$\mathfrak{e}$	$\mathfrak{e}$

$\neg$	
$\mathfrak{f}$	$\mathfrak{t}$
$\mathfrak{t}$	$\mathfrak{f}$
$\mathfrak{e}$	$\mathfrak{e}$

A formula φ entails a formula ψ in K_3^w if, for all valuations v, if $v(\varphi) = \mathfrak{t}$ then $v(\psi) = \mathfrak{t}$. Similarly, φ entails ψ in PWK if, for all valuations v, if $v(\psi) = \mathfrak{f}$ then $v(\varphi) = \mathfrak{f}$.

Theorem 1 *For all formulas φ and ψ, we have $\varphi \models_s \psi$ if and only if φ entails ψ in K_3^w.*

Theorem 2 *For all formulas φ and ψ, we have $\varphi \models_t \psi$ if and only if φ entails ψ in PWK.*

The truth-value $\mathfrak{e}$ in the Weak Kleene truth tables is what we call infectious: if $\mathfrak{e}$ appears as an argument of one of the operations interpreting one of the logical connectives, it is automatically the output of that operation.

---

[4] As noted by an anonymous reviewer, the ordering of possibilist frames does not play any role in the semantics for conjunction, disjunction and negation. Consequently, the semantics could be simplified by removing the ordering from the frames and the induced logic would be the same. However, our goal here is not to find the simplest semantics for a particular logic. Our goal, as mentioned in the introduction, is to see what kind of logic is induced by a particular notion of propositional content. Accordingly, our model theory is not a mere formal machinery used to characterise a consequence relation. It aims to be able to represent propositional contents and the ways they are interconnected. To this effect, the ordering appears crucial since the philosophical motivations of the model theory starts by taking states to be partial. Including the ordering in possibilist frames allows us to describe how that partiality interacts with our notions of truthmaking and falsitymaking.

Truthmaker Semantics for Infectious Logics

As such, K_3^w and PWK are called infectious logics (e.g., Ciuni, Szmuc, & Ferguson, 2018). In fact, they are the paradigmatic examples of infectious logics. The questions we would like to answer now are whether the semantics we just presented can be used to capture other infectious logics and how we could interpret those results philosophically.

4 Infectious logics

Before we can answer these questions, we need to introduce formally a wide range of infectious logics: the truth-functional infectious logics. To do so, we use the framework of matrix semantics. Let us introduce a few standard definitions.

A vocabulary is a pair $\Sigma = \langle V, n \rangle$ where V is a non-empty set and $n : V \to \mathbb{N} \setminus \{0\}$. The elements of V are called logical connectives and for each logical connective $\dagger$, the number $n(\dagger)$ is its arity. The language $\mathcal{L}_\Sigma$ is defined by the set of formulas generated by the following BNF:

$$\varphi ::= p \in Prop \mid \dagger(\varphi_1, ..., \varphi_{n(\dagger)}) \text{ for } \dagger \in V$$

A logical algebra $\mathcal{A}$ (for the vocabulary Σ) is a pair $\langle A, (.)^{\mathcal{A}} \rangle$ where A is a non-empty set of truth-values and $(.)^{\mathcal{A}}$ is a function which associates every $\dagger \in V$ with a function $\dagger^{\mathcal{A}} : A^{n(\dagger)} \to A$. Given such a logical algebra $\mathcal{A}$, we define a $\mathcal{A}$-valuation as a function $v : Prop \to A$. Every $\mathcal{A}$-valuation v can be canonically extended to a function $v : \mathcal{L}_\Sigma \to A$ by recursively stipulating that, for all $\dagger \in V$ and for all $\varphi_1, ..., \varphi_{n(\dagger)}$ in $\mathcal{L}_\Sigma$, we have $v(\dagger(\varphi_1, ..., \varphi_{n(\dagger)})) = \dagger^{\mathcal{A}}(v(\varphi_1), ..., v(\varphi_{n(\dagger)}))$. A logical matrix $\mathcal{M}$ (for the vocabulary Σ) is a tuple $\langle \mathcal{A}, D \rangle$ where $\mathcal{A}$ is a logical algebra and D is a non-empty proper subset of A of designated truth-values. Given a logical matrix $\mathcal{M} = \langle \mathcal{A}, D \rangle$, we say that a formula φ of $\mathcal{L}_\Sigma$ entails a formula ψ of $\mathcal{L}_\Sigma$ in $\mathcal{M}$, written $\varphi \models^{\mathcal{M}} \psi$, if if $v(\varphi) \in D$ entails $v(\psi) \in D$ for all $\mathcal{A}$-valuations v.

Let $\mathcal{WK}$ be the logical algebra induced by the Weak Kleene truth tables. The logic K_3^w coincides with the logic of the matrix $\langle \mathcal{WK}, \{t\} \rangle$ and the logic of PWK coincides with the logic of the matrix $\langle \mathcal{WK}, \{t, e\} \rangle$.

More generally, if $\mathcal{A}$ is a logical algebra, we can consider its infectious extension $\mathcal{A}^e$ defined as $\langle A \cup \{e\}, (.)^{\mathcal{A}^e} \rangle$ where, for all $\dagger \in V$ and $a_1, ..., a_{n(\dagger)} \in A \cup \{e\}$, we have $\dagger^{\mathcal{A}^e}(a_1, ..., a_{n(\dagger)}) = e$ if $e \in \{a_1, ..., a_{n(\dagger)}\}$ and $\dagger^{\mathcal{A}}(a_1, ..., a_{n(\dagger)})$ otherwise. One notices that e is infectious in $\mathcal{A}^e$. Moreover, the logical algebra $\mathcal{WK}$ is the infectious extension of the two-valued logical matrix $\mathcal{C}$ of classical logic, i.e., $\mathcal{WK} = \mathcal{C}^{\mathfrak{e}}$.

Every logical matrix $\mathcal{M} = \langle \mathcal{A}, D \rangle$ has two infectious extensions: its strict infectious extension $\mathcal{M}^s = \langle \mathcal{A}^{\mathfrak{e}}, D \rangle$ and its tolerant infectious extension $\mathcal{M}^t = \langle \mathcal{A}^{\mathfrak{e}}, D \cup \{\mathfrak{e}\} \rangle$. The difference between the two consists in the designation status of $\mathfrak{e}$. Notice that K_3^w is the logic of the strict infectious extension of classical logic while PWK is the logic of the tolerant infectious extension of classical logic.

The goal now is to devise, given a logical matrix $\mathcal{M}$, a semantics for the logic of $\mathcal{M}^s$ and $\mathcal{M}^t$. To do so, one can observe that truthmaking and falsitymaking clauses of the truthmaker semantics presented in the previous section are closely linked to the truth tables of classical logic. In other words, they are linked to $\mathcal{C}$. Indeed, to focus on the new disjunction clause, a state s makes $\varphi \vee \psi$ true if and only if s gives the truth-value $x \in \{\mathfrak{t}, \mathfrak{f}\}$ to φ, s gives the truth-value $y \in \{\mathfrak{t}, \mathfrak{f}\}$ to ψ and $x \vee^{\mathcal{C}} y = \mathfrak{t}$. They are three cases: $x = y = \mathfrak{t}$, $x = \mathfrak{t}$ and $y = \mathfrak{f}$, or $x = \mathfrak{f}$ and $y = \mathfrak{t}$. We get back our three possibilities for a true disjunction. Similar results hold for the truthmaking and falsitymaking clauses of all the classical connectives.

The question now is how to generalise to an arbitrary vocabulary Σ and an arbitrary logical matrix $\mathcal{M} = \langle \mathcal{A}, D \rangle$ in that vocabulary. In $\mathcal{C}$ there are two truth-values, $\mathfrak{t}$ and $\mathfrak{f}$, and our previous semantics contains two semantic relations, truthmaking ($\Vdash^+$) and falsitymaking ($\Vdash^-$). In $\mathcal{M}$, by contrast, there are $|A|$ truth-values. Therefore, in the semantics for $\mathcal{M}$, there should be one semantic relation for each truth-value $a \in A$, namely a-making (written $\Vdash^a$). The possibilist models for $\mathcal{M}$ should therefore contain an evaluation function v^a for all $a \in A$. Obviously, they should all respect the conditions that v^+ and v^- respected. Correspondingly, every connective $\dagger \in V$ should have a a-making clause for all $a \in A$. Generalising from the previous semantics, we get that a state s a-makes $\dagger(\varphi_1, ..., \varphi_{n(\dagger)})$ if and only if, for some $a_1, ..., a_{n(\dagger)} \in A$, we have that s a_1-makes φ_1, ..., s $a_{n(\dagger)}$-makes $\varphi_{n(\dagger)}$ and $\dagger^{\mathcal{A}}(a_1, ..., a_{n(\dagger)}) = a$.

Let's get formal. A possibilist model for $\mathcal{M}$ is a tuple $\langle S, \sqsubseteq, \{v^a \mid a \in A\} \rangle$ where $\langle S, \sqsubseteq \rangle$ is a possibilist frame and, for all $a \in A$, we have $v^a : Prop \to \mathcal{P}(S)$ such that, for all $p \in Prop$ and $s, t \in S$, (a) if $s \in v^a(p)$ for some $a \in A$ and $s \sqsubseteq t$ then $t \in v^a(p)$ and (b) if $s \in v^a(p)$ and $t \in v^{a'}(p)$ for some distinct $a, a' \in A$ then s and t are incompatible. Again, the first condition ensures that a-making is preserved by mereological extension and the second condition rules out inconsistent states.

We can now define, for all $a \in A$, the a-making relation for all formulas of $\mathcal{L}_\Sigma$:

- $s \Vdash^a p$ iff $s \in v^a(p)$,
- $s \Vdash^a \dagger(\varphi_1, ..., \varphi_{n(\dagger)})$ iff, for some $a_1, ..., a_{n(\dagger)} \in A$, we have that $s \Vdash^{a_1} \varphi_1, ..., s \Vdash^{a_{n(\dagger)}} \varphi_{n(\dagger)}$ and $\dagger^{\mathcal{A}}(a_1, ..., a_{n(\dagger)}) = a$.

We check that the conditions on v^a also holds for $\Vdash^a$.

Proposition 3 *For all $a \in A$, formulas φ and states $s, t \in S$ such that $s \sqsubseteq t$, if $s \Vdash^a \varphi$ then $t \Vdash^a \varphi$.*

Proposition 4 *For all formulas φ and states $s, t \in S$, if $s \Vdash^a \varphi$ and $t \Vdash^{a'} \varphi$ for distinct $a, a' \in A$ then s and t are incompatible.*

We say that a state s designates a formula φ if $s \Vdash^a \varphi$ for some $a \in D$. Similarly, we say that s anti-designates φ if $s \Vdash^a \varphi$ for some $a \in A \setminus D$. Here again it is possible to define two notions of logical consequence, depending on whether one cares about preservation of designatedness or preservation of non-anti-designatedness. A formula φ strictly entails a formula ψ, written $\varphi \vDash_s^{\mathcal{M}} \psi$, if every state which designates φ also designates ψ in all possibilist models for $\mathcal{M}$. By contrast, φ tolerantly entails ψ, written $\varphi \vDash_t^{\mathcal{M}} \psi$, if every state which anti-designates ψ also anti-designates φ in all possibilist models for $\mathcal{M}$.

As desired, the logic of strict entailment in the semantics for $\mathcal{M}$ coincides with the logic of the strict infectious extension of $\mathcal{M}$. Similarly, the logic of the tolerant entailment in the semantics for $\mathcal{M}$ coincides with the logic of the tolerant infectious extension of $\mathcal{M}$.

Theorem 3 *For all logical matrix $\mathcal{M}$ and formulas φ and ψ, we have $\varphi \vDash_s^{\mathcal{M}} \psi$ if and only if $\varphi \vDash^{\mathcal{M}^s} \psi$.*

Theorem 4 *For all logical matrix $\mathcal{M}$ and formulas φ and ψ, we have $\varphi \vDash_t^{\mathcal{M}} \psi$ if and only if $\varphi \vDash^{\mathcal{M}^t} \psi$.*

One notices that the truthmaker semantics presented in the previous section is exactly the semantic for $\mathcal{C}$. The Theorems 1 and 2 are therefore instances of the Theorems 3 and 4, respectively, where $\mathcal{M} = \mathcal{C}$.[5]

---

[5]This semantics is in a certain respects quite close to (Priest, 2014)'s plurivalent semantics and in particular to its singular version in (Szmuc & Omori, 2018). This is not a coincidence. In fact, one can prove the equivalence between plurivalent semantics and the version of our truthmaker semantics which includes impossible states. The connection between our framework and plurivalent semantics will be explored in future works.

5 Discussion

The infectious element e in the Weak Kleene truth tables, and more generally in infectious extensions of any logical algebra, has been given several interpretations.[6] We can mention (Fitting, 1994) and (Szmuc, 2019)'s epistemic interpretation, (Ferguson, 2014)'s computational interpretation and (Beall, 2016)'s off-topic interpretation. The original interpretation from (Bochvar & Bergmann, 1981) and (Halldén, 1949), however, is the nonsense interpretation according to which e is to represent meaninglessness. Under this interpretation, infectious logics are called nonense logics.

The general idea of nonsense logics starts from the recognition that grammatically correct sentences might not be truth-apt or even falsity-apt. A sentence like "the number 3 is green" is clearly not true but it is also not false, it is meaningless. Category mistakes but also sentences that are not correctly typed (e.g., under Russell's type theory) would be examples of such meaningless sentences. But these sentences can occur in reasoning and should therefore be dealt with by our logical theories. The question is then: how meaningless sentences interact with logical connectives? The fundamental principle used by nonsense logics has been called the *Principle of Component Homogeneity* by Goddard and Routley (1973) and states that any complex sentence which has a meaningless component is itself meaningless. In a framework in which meaninglessness is itself a truth-value in a matrix semantics, the Principle of Component Homogeneity demands that meaninglessness is infectious.[7] This grounds the interpretation of infectious logics as nonsense logics.

The results presented in this paper suggests an alternative interpretation of the infectious truth-value. Indeed, in our truthmaker semantics for infectious logics, the presence of e corresponds to the fact that states are partial and can be silent on the truth-value of a proposition. Assuming our new treatment of disjunction, the fact that a state does not determine the truth-value of the proposition "it is raining" entails the fact that this state does not determine the truth-value of "it is raining or it is sunny". But this is not because "it is raining" is meaningless and infects "it is raining or it is sunny". Rather, it is because this state's silence about rain entails its silence about rain and sun. Silence is infectious. Nothing here prevents the sentence "it is raining" to have a meaning, for instance if it is made true or false by other states. The

---
[6]For an overview, see (Omori & Szmuc, 2017).
[7]See (Szmuc & Omori, 2018) for a discussion.

partiality interpretation of e is then the following: a valuation attributes e to a proposition if that valuation represents a partial state which is silent about that proposition. The infectiousness of e represent then the infectiousness of silence. One might notice that what we call silence can also be called a truth-value gap, the absence of truth-value. The idea that truth-value gaps act infectiously is quite non-standard but has been advanced in (Da Ré, Pailos, & Szmuc, 2020) and applied to develop theories of truth. Of course, a full evaluation of the partiality interpretation is beyond the scope of this paper, whose aim was to present the formal results grounding its plausibility.

6 Conclusion

In this paper, we have taken seriously the idea that a truthmaker is a particular way for a proposition to be true. We aimed to study the formal consequences of a conception of ways to be true according to which they are exhaustive but also exclusive of the truth of the proposition in question. This led us towards a new treatment of disjunction and ultimately to a sound and complete inexact truthmaker semantics for Weak Kleene Logic and Paraconsistent Weak Kleene Logic, the infectious extensions of classical logic. We then saw that our semantics can be generalised to capture the infectious extensions of every truth-functional logic of any propositional language. This allowed to put forward a new interpretation of infectious truth-values: the partiality interpretation. More generally, the formal results presented in this paper cast a new light on infectious logics, and more particularly on Weak Kleene Logic and Paraconsistent Weak Kleene Logic. Indeed, they can now be motivated from a truth-conditional perspective through a particular conception of propositional content.

Throughout the paper, we have left open some avenues for future work. First, we have only considered possible and consistent partial situations in our truthmaker semantics. Our next step would then be to allow for the partial situations to be impossible and inconsistent. What kind of logics one then gets? A specific study should be dedicated to this question. Second, we only have considered the simultaneity reading of the new clause for disjunction. One might wonder what kind of semantics one gets with the mereological reading. This would then lead to a new type of exact truthmaker semantics. Note that here as well one should consider separately a framework with only possible and consistent partial states and a framework which also allows impossible and inconsistent states. Third and last, a thorough philosophical evaluation of the partiality interpretation of infectiousness still needs to be realised, for we only suggested its plausibility. It stills needs to be compared more closely with the nonsense interpretation, as we sketched above, but also with the other interpretations of infectiousness present in the literature.

Appendix

We give the proofs of the Proposition 3 and 4 and of the Theorems 3 and 4. The Propositions 1 and 2 and the Theorems 1 and 2 follow as corollaries.

Proof of Proposition 3. We proceed by induction on the complexity of φ.

If $\varphi = p$ for some $p \in Prop$, the result follows from the definition of v^a.

Suppose $\varphi = \dagger(\varphi_1, ..., \varphi_{n(\dagger)})$ for some $\dagger \in V$ and formulas $\varphi_1, ..., \varphi_{n(\dagger)}$ of $\mathcal{L}_\Sigma$. Since $s \Vdash^a \varphi$, we know that there are some $a_1, ..., a_{n(\dagger)} \in A$, such that $s \Vdash^{a_1} \varphi_1, ..., s \Vdash^{a_{n(\dagger)}} \varphi_{n(\dagger)}$ and $\dagger^A(a_1, ..., a_{n(\dagger)}) = a$. Using the induction hypothesis, we get that $t \Vdash^{a_1} \varphi_1, ..., t \Vdash^{a_{n(\dagger)}} \varphi_{n(\dagger)}$. It follows that $t \Vdash^a \varphi$, as desired. $\square$

Proof of Proposition 4. We proceed by induction on the complexity of φ.

If $\varphi = p$ for some $p \in Prop$, the result follows from the definition of v^a.

Suppose $\varphi = \dagger(\varphi_1, ..., \varphi_{n(\dagger)})$ for some $\dagger \in V$ and formulas $\varphi_1, ..., \varphi_{n(\dagger)}$ of $\mathcal{L}_\Sigma$. Since $s \Vdash^a \varphi$, we know that there are some $a_1, ..., a_{n(\dagger)} \in A$, such that $s \Vdash^{a_1} \varphi_1, ..., s \Vdash^{a_{n(\dagger)}} \varphi_{n(\dagger)}$ and $\dagger^A(a_1, ..., a_{n(\dagger)}) = a$. Similarly, since $t \Vdash^{a'} \varphi$, we know that there are some $a'_1, ..., a'_{n(\dagger)} \in A$, such that $t \Vdash^{a'_1} \varphi_1, ..., t \Vdash^{a'_{n(\dagger)}} \varphi_{n(\dagger)}$ and $\dagger^A(a'_1, ..., a'_{n(\dagger)}) = a'$. We know that $a \neq a'$ and therefore that there must be some i between 1 and $n(\dagger)$ such that $a_1 \neq a'_i$. Using the induction hypothesis, we deduce that s and t are incompatible, as desired. $\square$

Before giving the proof of Theorems 3 and 4, we need to do a little of work. For all states s, we define a $\mathcal{A}^{\mathfrak{e}}$-valuation as follows:

$$v_s : p \mapsto \begin{cases} a \in A \text{ if } s \in v^a(p) \\ \mathfrak{e} \text{ if there is no } a \in A \text{ such that } s \in v^a(p) \end{cases}$$

As a function, v_s is well-defined because, by definition of possibilist model for $\mathcal{M}$, there cannot be two distinct $a, a' \in A$ such that $s \in v^a(p)$ and $s \in v^{a'}(p)$.

Truthmaker Semantics for Infectious Logics

Lemma 1 *The extension of v_s to $\mathcal{L}_\Sigma$ is given as follows:*

$$v_s : \varphi \mapsto \begin{cases} a \in A \text{ if } s \Vdash^a \varphi \\ \mathfrak{e} \text{ if there is no } a \in A \text{ such that } s \Vdash^a \varphi \end{cases}$$

Proof. We start by noticing that this a well-defined function because, by Proposition 4, there cannot be two distinct $a, a' \in A$ such that $s \Vdash^a \varphi$ and $s \Vdash^{a'} \varphi$.

We proceed by induction on the complexity of φ.

If $\varphi = p$ for some $p \in Prop$, the result follows from the definition of v_s.

Suppose $\varphi = \dagger(\varphi_1, ..., \varphi_{n(\dagger)})$ for some $\dagger \in V$ and formulas $\varphi_1, ..., \varphi_{n(\dagger)}$ of $\mathcal{L}_\Sigma$.

If $s \Vdash^a \varphi$ for some $a \in A$, then we know that there are some $a_1, ..., a_{n(\dagger)} \in A$, such that $s \Vdash^{a_1} \varphi_1, ..., s \Vdash^{a_{n(\dagger)}} \varphi_{n(\dagger)}$ and $\dagger^{\mathcal{A}}(a_1, ..., a_{n(\dagger)}) = a$. By induction hypothesis, it follows that $v_s(\varphi_1) = a_1, ..., v_s(\varphi_{n(\dagger)}) = a_{n(\dagger)}$. Thus, $v_s(\varphi) = \dagger^{\mathcal{A}}(v_s(\varphi_1), ..., v_s(\varphi_{n(\dagger)})) = \dagger^{\mathcal{A}}(a_1, ..., a_{n(\dagger)}) = a$.

If, however, there is no $a \in A$ such that $s \Vdash_a \varphi$, then there must be some i between 1 and $n(\dagger)$ such that there is no $a \in A$ such that $s \Vdash^a \varphi_i$. By induction hypothesis, we have $v_s(\varphi_i) = \mathfrak{e}$ and therefore $v_s(\varphi) = \mathfrak{e}$. □

Proposition 5 *For all formulas φ and ψ of $\mathcal{L}_\Sigma$, if $\varphi \models^{\mathcal{M}_s} \psi$ then $\varphi \models^{\mathcal{M}}_s \psi$.*

Proof. Let $\langle S, \sqsubseteq, \{v_a \mid a \in A\} \rangle$ be a possibilist model for $\mathcal{M}$ and let $s \in S$ such that s designates φ. So there is some $a \in D$ such that $s \Vdash^a \varphi$. So $v_s(\varphi) = a$ and thus $v_s(\varphi) \in D$. Since $\varphi \models_{\mathcal{M}_s} \psi$, it follows that $v_s(\psi) \in D$. So there is some $a' \in D$ such that $v_s(\psi) = a'$. So $s \Vdash^{a'} \psi$. Since $a' \in D'$, it means that s designates ψ. Consequently, we have $\varphi \models^{\mathcal{M}}_s \psi$. □

Proposition 6 *For all formulas φ and ψ of $\mathcal{L}_\Sigma$, if $\varphi \models^{\mathcal{M}_t} \psi$ then $\varphi \models^{\mathcal{M}}_t \psi$.*

Proof. Let $\langle S, \sqsubseteq, \{v_a \mid a \in A\} \rangle$ be a possibilist model for $\mathcal{M}$ and let $s \in S$ such that s anti-designates ψ. So there is some $a \in A \setminus D$ such that $s \Vdash^a \psi$. Assume, for reductio, that s does not anti-designate φ. So either $s \Vdash^{a'} \varphi$ for some $a' \in D$ or there is no $a' \in A$ such that $s \Vdash^{a'} \varphi$. Consequently, $v_s(\varphi) \in D$ or $v_s(\varphi) = \mathfrak{e}$. Since $v_s(\varphi) \in D \cup \{\mathfrak{e}\}$ and $\varphi \models_{\mathcal{M}_t} \psi$, it follows that $v_s(\psi) \in D \cup \{\mathfrak{e}\}$, which contradicts $a \in A \setminus D$. Thus, s anti-designates φ. Therefore, we have $\varphi \models^{\mathcal{M}}_t \psi$. □

We now build a canonical possibilist model for $\mathcal{M}$.[8] Let S_c be the set of $\mathcal{A}^{\mathfrak{e}}$-valuations. If $v, v' \in S_c$, we write $s \sqsubseteq_c s'$ if, for all $a \in A$ and $p \in Prop$, if $v(p) = a$ then $v'(p) = a$. Clearly, $\sqsubseteq_c$ is a partial order. Now, for $a \in A$ and $Prop$, we define $v_c^a(p) = \{v \in S_c \,|\, v(p) = a\}$.

Lemma 2 *The structure $\langle S_c, \sqsubseteq_c, \{v_c^a \,|\, a \in A\}\rangle$ is a possibilist model.*

Proof. We first check that, for $a \in A$ and $s, t \in S_c$, if $s \in v_c^a(p)$ and $s \sqsubseteq_c t$ then $t \in v_c^a(p)$. We have $s \in v_c^a(p)$ so $s(p) = a$. Since $s \sqsubseteq_c t$, we get $t(p) = a$ and so $t \in v_c^a(p)$.

Second, we check that, for $s, t \in S_c$ and distinct $a, a' \in A$, if $s \in v_c^a(p)$ and $t \in v_c^{a'}(p)$ then s and t are incompatible. Suppose not. Then there is some $u \in S_c$ such that $s \sqsubseteq_c u$ and $t \sqsubseteq_c u$. So $u(p) = a$ and $u(p) = a'$, which is a contradiction. $\square$

Lemma 3 *For all $s \in S_c$, formula φ and $a \in A$, we have $s \Vdash^a \varphi$ if and only if $s(\varphi) = a$.*

Proof. By definition, we have $v_s = s$. The result then follows from Lemma 1. $\square$

Proposition 7 *For all formulas φ and ψ of $\mathcal{L}_\Sigma$, if $\varphi \models_s^{\mathcal{M}} \psi$ then $\varphi \models^{\mathcal{M}_s} \psi$.*

Proof. Let v be a $\mathcal{A}^{\mathfrak{e}}$-valuation such that $v(\varphi) \in D$. So $v(\varphi) = a$ for some $a \in A$. Thus, $v \Vdash^a \varphi$ in the canonical model. Since $a \in D$, v designates φ. Because $\varphi \models_s^{\mathcal{M}} \psi$, we know that v designates ψ. So there is some $a' \in D$ such that $v \Vdash^{a'} \psi$. Therefore, $v(\psi) = a'$. Consequently, $\varphi \models^{\mathcal{M}_s} \psi$. $\square$

Proposition 8 *For all formulas φ and ψ of $\mathcal{L}_\Sigma$, if $\varphi \models_t^{\mathcal{M}} \psi$ then $\varphi \models^{\mathcal{M}_t} \psi$.*

Proof. Let v be a $\mathcal{A}^{\mathfrak{e}}$-valuation such that $v(\varphi) \in D \cup \{\mathfrak{e}\}$. Assume, for reductio, that $v(\psi) \in A \setminus D$. So $v(\psi) = a$ for some $a \in A \setminus D$. Thus, $v \Vdash^a \psi$ in the canonical model. Since $a \notin D$, v anti-designates ψ. Because $\varphi \models_t^{\mathcal{M}} \psi$, we know that v anti-designates φ. So there is some $a' \in A \setminus D$ such that $v \Vdash^{a'} \varphi$. Therefore, $v(\varphi) = a'$ and so $v(\varphi) \in A \setminus D$, which is a contradiction. It follows that $v(\psi) \in D \cup \{\mathfrak{e}\}$. Consequently, $\varphi \models^{\mathcal{M}_t} \psi$. $\square$

---

[8] As noted by an anonymous reviewer, a simpler proof of completeness can be given by taking a model with only one state invalidating the particular inference we care about. We prefer to give the slightly more complicated proof which uses a model invalidating all invalid inferences. The reason is again linked to our motivations. Our model theory aims to represent a particular notion of propositional content, which in mind the idea that it will be possible to apply that notion to the analysis of a particular propositional attitudes. To this effect, it is crucial that a multiplicity of propositional contents can be represented within a single model. This is shown by the canonical construction we provide.

Proof of Theorem 3. It follows from Propositions 5 and 7. □

Proof of Theorem 4. It follows from Propositions 6 and 8. □

References

Barwise, J., & Perry, J. (1981). Situations and attitudes. *The Journal of Philosophy, 78*(11), 668–691.

Beall, J. (2016). Off-topic: A new interpretation of weak-Kleene logic. *The Australasian Journal of Logic, 13*(6).

Bochvar, D. A., & Bergmann, M. (1981). On a three-valued logical calculus and its application to the analysis of the paradoxes of the classical extended functional calculus. *History and Philosophy of Logic, 2*(1-2), 87–112.

Ciuni, R., & Carrara, M. (2019). Semantical analysis of Weak Kleene logics. *Journal of Applied Non-Classical Logics, 29*(1), 1–36.

Ciuni, R., Szmuc, D., & Ferguson, T. M. (2018). Relevant logics obeying component homogeneity. *The Australasian Journal of Logic, 15*(2), 301–361.

Da Ré, B., Pailos, F., & Szmuc, D. (2020). Theories of truth based on four-valued infectious logics. *Logic Journal of the IGPL, 28*(5), 712–746.

Deigan, M. (2020). A plea for inexact truthmaking. *Linguistics and Philosophy, 43*(5), 515–536.

Ferguson, T. (2014). A computational interpretation of conceptivism. *Journal of Applied Non-Classical Logics, 24*(4), 333–367.

Fine, K. (2016). Angellic content. *Journal of Philosophical Logic, 45*(2), 199–226.

Fine, K. (2017a). A theory of truthmaker content I: Conjunction, disjunction and negation. *Journal of Philosophical Logic, 46*(6), 625–674.

Fine, K. (2017b). Truthmaker semantics. In B. Hale, C. Wright, & A. Miller (Eds.), *A Companion to the Philosophy of Language* (2nd ed., pp. 556–577). Hoboken, NJ: John Wiley & Sons, Ltd.

Fitting, M. (1994). Kleene's three valued logics and their children. *Fundamenta Informaticae, 20*(1, 2, 3), 113–131.

Genco, F. A., Poggiolesi, F., & Rossi, L. (2021). Grounding, quantifiers, and paradoxes. *Journal of Philosophical Logic, 50*(6), 1417–1448.

Goddard, L., & Routley, R. (1973). *The Logic of Significance and Context*. Edinburgh: Scottish Academic Press.

Halldén, S. (1949). *The Logic of Nonsense*. Upsala: Upsala Universitets Arsskrift.

Hawke, P., & Özgün, A. (in press). Truthmaker semantics for epistemic logic. *Outstanding Contributions to Logic-Kit Fine*.

Jago, M. (2020). Truthmaker semantics for relevant logic. *Journal of Philosophical Logic*, 49(4), 681–702.

Korbmacher, J., Anglberger, A., & Faroldi, F. L. (2016). An exact truthmaker semantics for permission and obligation. In O. Roy, A. Tamminga, & M. Willer (Eds.), *Deontic Logic and Normative Systems. DEON 2016*. (pp. 16–31). London: College Publications.

Leitgeb, H. (2019). HYPE: A system of hyperintensional logic (with an application to semantic paradoxes). *Journal of Philosophical Logic*, 48(2), 305–405.

Leitgeb, H. (2021). Exact truthmaking as inexact truthmaking by minimal totality facts. In *Logic in High Definition* (pp. 67–75). Cham: Springer.

Omori, H., & Szmuc, D. (2017). Conjunction and disjunction in infectious logics. In A. Baltag, J. Seligman, & T. Yamada (Eds.), *Logic, Rationality, and Interaction. LORI 2017* (pp. 268–283). Berlin, Heidelberg.

Priest, G. (2014). Plurivalent logics. *Australasian Journal of Logic*, 11(1).

Szmuc, D. (2019). An epistemic interpretation of Paraconsistent Weak Kleene logic. *Logic and Logical Philosophy*, 28(2), 277–330.

Szmuc, D., & Omori, H. (2018). A note on Goddard and Routley's significance logic. *The Australasian Journal of Logic*, 15(2), 431–448.

Van Fraassen, B. C. (1969). Facts and tautological entailments. *The Journal of Philosophy*, 66(15), 477–487.

Yablo, S. (2014). *Aboutness*. Princeton: Princeton University Press.

Thomas Randriamahazaka
University of St Andrews
United Kingdom
E-mail: tr52@st-andrews.ac.uk

Decidability in Proof-Theoretic Validity

WILL STAFFORD[1]

Abstract: Proof-theoretic validity has proven a useful tool for proof-theoretic semantics, because it explains the harmony found in the introduction and elimination rules for the intuitionistic calculus. However, the demonstration that a rule of proof is proof-theoretically valid requires checking an infinite number of cases, which raises the question of whether proof-theoretic validity is decidable. It is proven here that it is for the most prominent formulations in the literature for propositional logic.

Keywords: Proof-theoretic validity, decidability, Prawitz

1 Introduction

Proof-theoretic validity is a property of proof-like structures which distinguishes those which are valid, and therefore proofs, from those which are not. It is used to define a consequence relation on formulas and so is an alternative to model-theoretic approaches. The idea was initially put forward by Prawitz (1971, 1973, 1974) but in the following, we work with the definition of proof-theoretically valid consequence found in Piecha, de Campos Sanz, and Schroeder-Heister (2015).

This paper is a response to a concern raised when I first started working on proof-theoretic validity. A proof-like structure, very roughly, is valid if it is built up of introduction rules or as one goes from proof to sub-proof and replaces assumptions with proof it can always be transformed into a proof built up of introduction rules. The worry is that the normal presentation of the definition of proof-theoretic validity requires the existence of a transformation for every possible proof of assumptions on the left of the consequence relation. This would be formalised as an unbounded existential quantifier followed

---

[1] I would like to thank Sean Walsh for discussions on this work, the anonymous reviewer for helpful comments, and the audiences of Logica 2021 and the 3rd Workshop on Proof Theory and its Applications for their questions.

by an unbounded universal. This implies that the definition as written is Σ_2 and thus not decidable. In this paper, we show that for privileged approaches to the atomic formulas, there is a decision procedure for propositional logic. We do this by demonstrating that these approaches have the finite model property and, as a consequence, any non-validity has a finite counter model.[2] This in essence puts a bound on the universal quantifier in the definition.

The paper is organised as follows: In Section 2, we introduce proof-theoretic validity for propositional logic, proof-theoretic systems, and atomic rules. In Section 3, we demonstrate when one can close a proof-theoretic system under derivable atomic rules and restrict a proof-theoretic system to certain atomic formulas without affecting what is valid. Finally, in Section 4, we define a set of decidable proof-theoretic systems that they have the finite model property.

2 Proof-theoretic validity

Let us start by outlining proof-theoretic validity. To do this, we must describe the treatment of atomic formulas.

2.1 Atomic rules

Because in this paper, we will work with proof-theoretic validity as a consequence relation, it can sometimes be hard to see what exactly is 'proof-theoretic' about this definition.[3] Still, the treatment of the atomic formulas is 'proof-theoretic'. Rather than assign a truth value to each atom, as we would in model-theoretic semantics, proof-theoretic validity considers a set of proof rules that contain only atomic formulas. We treat $\bot$ as simply another atomic formula for these purposes. Simple rules of this type might be axioms, such as $\bar{p}$, or inferences from atoms to atoms, such as $\frac{p\ q\ r}{s}$. We also need more complex rules of inference that can discharge inferences used earlier in the proof. These complex rules are considered despite not appearing in Prawitz's original formulation because they allow us to give proof-theoretic semantics to several intermediate logics extending Kriesel-Putnam logic (e.g. inquisitive logic; see Stafford (2021)). We introduce these rules via the following definition:

---

[2] The reader might be confused by the use of the word 'models' when proof-theoretic validity is supposed to be opposed to model-theoretic semantics but the models in this case are sets of proof rules for the atomic formulas.

[3] The definition is still equivalent to a definition on proofs as shown in Stafford (2021).

Decidability in Proof-Theoretic Validity

Definition 1 *(Schroeder-Heister 1984) We define atomic rule formulas and their levels as follows:*

> *(1.1.) A level-0 atomic rule is an axiom consisting of a single atomic formula. It is a rule with no premises or hypotheses. It is written as $\bar{p}$ or $/p$.*
>
> *(1.2.) A level-1 atomic rule is a rule that has premises but does not discharge hypotheses. It is written as $p_0,\ldots,p_n/q$ for an inference from $p_0,\ldots,p_n$ to q.*
>
> *(1.3.) A level-2 atomic rule is a rule which discharges hypotheses. It is written as*
>
> $$[p_{0_0},\ldots,p_{m_0}]q_0,\ldots,[p_{0_n},\ldots,p_{m_n}]q_n/r$$
>
> *for an inference from $q_0,\ldots,q_n$ to r, which for each q_i discharges $p_{0_i},\ldots,p_{m_i}$.*
>
> *(1.4.) A level-n atomic rule (for $n > 2$) is a rule which discharge rules of level-n-2. It is written as*
>
> $$([R_{0_0},\ldots,R_{m_0}]q_0),\ldots,([R_{0_n},\ldots,R_{m_n}]q_n)/r$$
>
> *for a level-n inference from $q_0,\ldots,q_n$ to r, which for each q_i discharges rules $R_{0_i},\ldots,R_{m_i}$, where the level of each R_j is at most level-n-2.*

A quick note about notational conventions: It will be helpful to write level-n+1 rules using level-n rules. For example, the rule $[p]q/r$ does not contain the rule p/q, but we could construct a function ^{R} which replaces $/$ with $[]$ and $[]$ with $/$. Then $[p]q/r \equiv (p/q)^R/r$. Similarly, $([s]t/u)^R, (/v)^R/r \equiv ([s/t]u), v/r$ and $(((s/t]u), v/r)^R/p \equiv [([s]t/u), v]r/p$. In what follows we will suppress the function ^{R} and simply write $R_1,\ldots,R_m/p$ to mean a rule generated by applying ^{R} to level-n rules $R_1,\ldots,R_m$ to generate a level-n+1 rule. Having defined the atomic rules we need to define the proof-theoretic equivalent of a model which we will call a proof-theoretic system.

Definition 2 *Let the set of all atomic rules of any level be denoted as $\mathbb{S}$.*

> *(2.1.) Call a set of atomic rules $S \subseteq \mathbb{S}$ an atomic system.*
>
> *(2.2.) Call a set $\mathfrak{S} \subseteq \mathcal{P}(\mathbb{S})$ a proof-theoretic system.*

We can see that a proof-theoretic system is made up of atomic systems. It may help the reader to think of a proof-theoretic system as being analogous to a Kripke model where each atomic system is analogous to a world in which one possible definition of the atomic formulas has been given and supersets of this atomic system represent possible extensions of these definitions.

Definition 3 *Let $S \vdash p$ if there is a proof of p using only rules in S. We will write $R_1, \ldots, R_n, S \vdash p$ for $\{R_1, \ldots, R_n\} \cup S \vdash p$.*

2.2 The consequence relation

Proof-theoretic validity can be defined on proof-like structures as a method of distinguishing those which are valid proofs from those which do not contain correct reasoning. It is fair to say that it is the definition of proofs that carries the philosophical weight. But, for technical purposes, we can work with a much more tractable definition provided by Piecha et al. (2015).

Definition 4 *For proof-theoretic systems $\mathfrak{S}$ and atomic systems $S \in \mathfrak{S}$, we define proof-theoretic validity $\models$ as follows:*

$$\mathfrak{S}, S \models p \iff S \vdash p, \tag{1}$$
$$\mathfrak{S}, S \models \varphi \land \psi \iff \mathfrak{S}, S \models \varphi \text{ and } \mathfrak{S}, S \models \psi, \tag{2}$$
$$\mathfrak{S}, S \models \varphi \lor \psi \iff \mathfrak{S}, S \models \varphi \text{ or } \mathfrak{S}, S \models \psi, \tag{3}$$
$$\mathfrak{S}, S \models \psi \to \varphi \iff [\forall S' \supseteq S(S' \in \mathfrak{S} \text{ and } \mathfrak{S}, S' \models \psi \Rightarrow \mathfrak{S}, S' \models \varphi)]. \tag{4}$$

Furthermore, the relation $\mathfrak{S} \models \varphi$ is defined as follows:

$$\mathfrak{S} \models \varphi \iff \forall S \in \mathfrak{S} : \mathfrak{S}, S \models \varphi. \tag{5}$$

This is the definition of proof-theoretic validity we will use throughout.

3 Closure and restriction

Recall that this paper shows that for certain proof-theoretic systems $\mathfrak{S}$, the question of whether a formula is valid on $\mathfrak{S}$ can be reduced to the question of whether it is valid on a finite proof-theoretic system. In aid of that, the goal of this section is to show that certain proof-theoretic systems can be modified without this affecting whether they are a counterexample to the validity of a formula. We will first show when we can close a proof-theoretic system under all derivable rules, and then describe when we can restrict a proof-theoretic system to rules containing only certain atomic letters.

Decidability in Proof-Theoretic Validity

3.1 Closure

The closure of a set of atomic rules is the set of all atomic rules that can be constructed from the initial rules. For example, if we had the set $\{p/q, q/r\}$, then we could construct the rule p/r by first going from p to q and then q to r. So, p/r is in the closure. The closure will ensure that we can restrict sets of atomic rules to certain atomic formulas. To see why the closure is necessary, note that if we do not make use of closure and restrict our example set to the atomic formulas p and r, we get the empty set. But this fails to preserve the relationship between p and r. With this in mind we define the closure as follows:

Definition 5 *For $S \subseteq \mathbb{S}$ define S^{cl} as S plus any rule $(R_0, \ldots, R_m)/p$ such that $S, R_0, \ldots, R_m \vdash p$ (where $\{R_0, \ldots, R_m\}$ can be empty).*

When we have $S, R_0, \ldots, R_m \vdash p$ we will say that S *witnesses* the rule $(R_0, \ldots, R_m)/p$. We want to demonstrate that for certain proof-theoretic systems closure does not change what is proof-theoretically valid. To do this we need to show that this is true for $\vdash$ from Definition 3. We can demonstrate this with the following result:

Theorem 1 *Given a proof $\mathcal{D}$ in $S \cup \{R\}$ and a proof $\mathcal{E}$ witnessing R in S', there is a proof constructed from $\mathcal{D}$ and $\mathcal{E}$ in $S \cup S'$ with the same conclusion as $\mathcal{D}$.*

Proof. The proof proceeds by induction on the level of the rule and the number of occurrences. Let R be the level-m rule we wish to remove. We can write R as $\frac{R_0 \ldots R_n}{q}$ where $R_1, \ldots, R_n$ are level-m-1 rules. For each R_j ($j \leq n$) let $R_{0_j}, \ldots, R_{u_j}$ be the level-m-2 rules (if any) which R discharges on that branch, such that $R_j \equiv (R_{0_j}, \ldots, R_{u_j})/p_j$. Let there be a witness $\mathcal{E}$ of $S', R_1, \ldots, R_n \vdash q$ and a proof $\mathcal{D}$ in $S \cup \{R\}$, where R occurs t times.

We have a core of $\mathcal{E}$, written $\mathcal{E}^C$ which is generated by snipping each branch in $\mathcal{E}$ at the first instance of one of the rules $R_1, \ldots, R_n$ that occurs above the root. This means that $\mathcal{E}^C$ is a proof from some collection of p_i for $i \in I$ to q and contains rules only in $S' \cup \{/p_i\}_{i \in I}$.

$$(\mathcal{E}_{p_i})_{i \in I}$$
$$(p_i)_{i \in I}$$
$$\mathcal{E}^C$$
$$q$$

We also have proofs $\mathcal{E}_{p_i}$ for $i \in I$ which are what remains of $\mathcal{E}$ after the core has been removed. Each proof $\mathcal{E}_{p_i}$ is in $S' \cup S^{p_i} \cup \{R_1,\ldots,R_n\}$, where S^{p_i} is all rules in $\mathcal{E}_{p_i}$ which are discharged in $\mathcal{E}^C$.

Take the proof $\mathcal{D}$ and identify an instance of R to remove, whereby there must not be any instances of R higher in the proof. Note that we can split $\mathcal{D}$ at this rule instance into the proof after the rule, $\mathcal{D}_q$, which is in $S \cup \{/q, R\}$ and proofs of each of the premises, $\mathcal{D}_{p_1},\ldots,\mathcal{D}_{p_n}$, which for each p_j is in $S \cup S^D \cup \{R_{1_j},\ldots,R_{u_j}\}$ where S^D is the set of all rules in $\mathcal{D}_{p_1},\ldots,\mathcal{D}_{p_n}$ which are discharged in $\mathcal{D}_q$. Also note that $\mathcal{D}_q$ has t-1 instances of R (all the other proofs have zero instances).

$$\frac{(\mathcal{D}_{p_i})_{i \in I}}{\begin{array}{c} (p_i)_{i \in I} \\ q \\ \mathcal{D}_q \end{array}} R$$

Consider that each $\mathcal{D}_{p_j}$ is a proof that $S \cup S^D, R_{1_j},\ldots,R_{u_j} \vdash p_j$, which means it is a witness for R_j in $S \cup S^D$. Also note that all R_js are of a lower level than R.

Take the proofs $\mathcal{E}_{p_i}$ for $i \in I$ (acquired by decomposing the witness) that are in $S' \cup S^{p_i} \cup \{R_1,\ldots,R_n\}$. We now have witnesses for $R_1,\ldots,R_n$, so we can remove these instances from each $\mathcal{E}_{p_i}$ by repeated application of the I.H. As a result, there is for each $\mathcal{E}_{p_i}$ a proof $\mathcal{D}^*_{p_i}$ in $S \cup S^D \cup S' \cup S^{p_i}$.

Glue the proofs together as follows:

$$\begin{array}{c} (\mathcal{D}^*_{p_i})_{i \in I} \\ (p_i)_{i \in I} \\ \mathcal{E}^C \\ q \\ \mathcal{D}_q \end{array}$$

This proof is in $S \cup S' \cup \{R\}$ because while the $\mathcal{D}^*_{p_i}$s are in $S \cup S^D \cup S' \cup S^{p_i}$, every rule in S^D is discharged in $\mathcal{D}_q$ and every rule in S^{p_i} is discharged in $\mathcal{E}^C$ (and placed on the correct branch to ensure that). Because only the $\mathcal{D}_q$ contained instances of R, there is one less instance of R. Therefore, by the induction hypothesis there exists a proof with the required properties. □

From this, it follows immediately that:

Decidability in Proof-Theoretic Validity

Corollary 1 $S \vdash p \Leftrightarrow S^{cl} \vdash p$

To extend this result we define the following property of proof-theoretic systems:

Definition 6 *We will call a proof-theoretic system $\mathfrak{S}$ closure ready if for $S', S^{cl} \in \mathfrak{S}^{cl}$ and S' extending S^{cl}, there is an $S'' \in \mathfrak{S}$ such that S'' extends S and $S' = (S'')^{cl}$.*

This property allows us to move back and forth between a proof-theoretic system and its closure, and it can be represented as follows:

$$
\begin{array}{ccc}
(\mathfrak{S}^{cl}) & & (\mathfrak{S}) \\
(S'')^{cl} = S' & & S'' \\
\uparrow & \Longrightarrow & \uparrow \\
S^{cl} & \Longleftarrow & S
\end{array}
$$

We also need the following result:

Lemma 1 $(S^{cl})^{cl} = S^{cl}$

Proof. Assume $R_1, \ldots, R_n/p \in (S^{cl})^{cl}$. It follows that $R_1, \ldots, R_n, S^{cl} \vdash p$ from which it follows by Corollary 1 that $R_1, \ldots, R_n, S \vdash p$, and thus that $R_1, \ldots, R_n/p \in S^{cl}$. □

We can now prove the following theorem:

Theorem 2 *For closure ready $\mathfrak{S}$, closure changes nothing:*

$$\mathfrak{S}, S \vDash \varphi \Leftrightarrow \mathfrak{S}^{cl}, S^{cl} \vDash \varphi.$$

Proof. The proof proceeds by induction on φ. The base case is Corollary 1. The only inductive case of interest is $\rightarrow$. Let $\mathfrak{S}, S \vDash \varphi \rightarrow \psi$. By the conditions for $\rightarrow$, it follows that for all $S' \in \mathfrak{S}$ that extend S if $\mathfrak{S}, S' \vDash \varphi$ the $\mathfrak{S}, S' \vDash \psi$.

Let $S' \in \mathfrak{S}^{cl}$ extend S^{cl} and $\mathfrak{S}^{cl}, S' \vDash \varphi$. Note that by closure readiness there is an $S'' \in \mathfrak{S}$ which extends S and $S'' = (S')^{cl}$. By two applications of the induction hypothesis, we thus arrive at $\mathfrak{S}^{cl}, S'' \vDash \psi$.

Let $\mathfrak{S}^{cl}, S^{cl} \vDash \varphi \rightarrow \psi$. By the conditions for $\rightarrow$, it follows that for all $S' \in \mathfrak{S}^{cl}$ that extend S^{cl}, if $\mathfrak{S}^{cl}, S' \vDash \varphi$, the $\mathfrak{S}^{cl}, S' \vDash \psi$. Let $S'' \in \mathfrak{S}$ extend S and $\mathfrak{S}, S'' \vDash \varphi$. Then $(S'')^{cl}$ extends S^{cl}. By two applications of the induction hypothesis, we thus arrive at $\mathfrak{S}, S'' \vDash \psi$. □

3.2 Restriction

In this section we will define a function on atomic systems and proof-theoretic systems that produces a new atomic system or proof-theoretic system with all rules restricted to atomic formulas. We will use the word 'restriction' to describe this. We will then show that given the closure of a atomic system or proof-theoretic system, this restriction preserves the property of being a counter-model. First, let us define restriction to a set as follows:

Definition 7 *Let* $\mathbb{S}^{\{p_0,\ldots,p_n\}}$ *be all rules R containing only atomic letters $p_0, \ldots, p_n$. Then let* $S^{\{p_0,\ldots,p_n\}} = S \cap \mathbb{S}^{\{p_0,\ldots,p_n\}}$ *and* $\mathfrak{S}^{\{p_0,\ldots,p_n\}} = \{S \cap \mathbb{S}^{\{p_0,\ldots,p_n\}} \mid S \in \mathfrak{S}\}$.

Just as with closure, we define the following property which describes the ability to go back and forth between the closure and its restriction.

Definition 8 *We will call a proof-theoretic system $\mathfrak{S}$ restriction ready if when $S', S^{\{p_0,\ldots,p_n\}} \in (\mathfrak{S}^{cl})^{\{p_0,\ldots,p_n\}}$ and S' extends $S^{\{p_0,\ldots,p_n\}}$, then there is an $S'' \in \mathfrak{S}^{cl}$ such that S'' extends S and $S' = (S'')^{\{p_0,\ldots,p_n\}}$.*

This now allows us to prove the following result:

Theorem 3 *For closure and restriction ready $\mathfrak{S}$ and φ containing only $p_1 \ldots p_n$, we have*

$$\mathfrak{S}^{cl}, S \vDash \varphi \Leftrightarrow (\mathfrak{S}^{cl})^{\{p_0,\ldots,p_n\}}, S^{\{p_0,\ldots,p_n\}} \vDash \varphi.$$

Proof. Proof by induction on φ. For the base case note that if $\mathfrak{S}^{cl}, S \vDash p$, then $/p \in S$. Let $\mathfrak{S}^{cl}, S \vDash \varphi \to \psi$. By the conditions for $\to$, it follows that for all $S' \in \mathfrak{S}^{cl}$ that extend S, if $\mathfrak{S}^{cl}, S' \vDash \varphi$ then $\mathfrak{S}^{cl}, S' \vDash \psi$. Let $S'' \in (\mathfrak{S}^{cl})^{\{p_0,\ldots,p_n\}}$ extend $S^{\{p_0,\ldots,p_n\}}$ and $(\mathfrak{S}^{cl})^{\{p_0,\ldots,p_n\}}, S'' \vDash \varphi$. Note that by restriction readiness, there is an $S'' \in \mathfrak{S}$ which extends S and $S'' = (S')^{\{p_0,\ldots,p_n\}}$. By two applications of the induction hypothesis, we thus arrive at $(\mathfrak{S}^{cl})^{\{p_0,\ldots,p_n\}}, S'' \vDash \psi$.

Let $(\mathfrak{S}^{cl})^{\{p_0,\ldots,p_n\}}, S^{\{p_0,\ldots,p_n\}} \vDash \varphi \to \psi$. By the conditions for $\to$, it follows that for all $S' \in (\mathfrak{S}^{cl})^{\{p_0,\ldots,p_n\}}$ that extend $S^{\{p_0,\ldots,p_n\}}$, if it is the case that $(\mathfrak{S}^{cl})^{\{p_0,\ldots,p_n\}}, S' \vDash \varphi$, then $(\mathfrak{S}^{cl})^{\{p_0,\ldots,p_n\}}, S' \vDash \psi$. Let $S'' \in \mathfrak{S}^{cl}$ extend S and $\mathfrak{S}^{cl}, S'' \vDash \varphi$. Then $(S'')^{\{p_0,\ldots,p_n\}}$ extends $S^{\{p_0,\ldots,p_n\}}$. By two applications of the induction hypothesis, we arrive at $\mathfrak{S}, S'' \vDash \psi$. □

Decidability in Proof-Theoretic Validity

3.3 Closure and restriction ready proof-theoretic systems

These results do not as yet deliver much because we do not know which systems are closure ready and/or restriction ready. To start with, let us take a closer look at how the two interact.

Lemma 2 *If $\mathfrak{S}$ and $\mathfrak{S}'$ are both closure ready or both restriction ready, then $\mathfrak{S} \cup \mathfrak{S}'$ is likewise.*

Proof. Let $S^{cl}, S' \in (\mathfrak{S} \cup \mathfrak{S}')^{cl}$ and let S' extend S^{cl}. We know that S' is the closure of a set in either $\mathfrak{S}$ or $\mathfrak{S}'$, therefore $S' \in \mathfrak{S}^{cl}$ or $S' \in \mathfrak{S}'^{cl}$. In either case, by closure readiness there is an S'' such that $S' = (S'')^{cl}$ and $S'' \in \mathfrak{S} \cup \mathfrak{S}'$. The proof for restriction readiness follows the same reasoning. □

Lemma 3 *If $\mathfrak{S}$ is closure ready or restriction ready, then any filter on $\mathfrak{S}$ is likewise closure or restriction ready.*

Proof. Let $\mathfrak{S}'$ be a filter on $\mathfrak{S}$ and let $S', S^{cl} \in \mathfrak{S}'^{cl}$ with S' extending S^{cl}. Then $S' \in \mathfrak{S}^{cl}$, and by the closure readiness of $\mathfrak{S}$, there is an $S'' \in \mathfrak{S}$ such that $S' = (S'')^{cl}$. Further, $S'' \in \mathfrak{S}'$ because filters are closed under supersets and $S \in \mathfrak{S}'$. The proof for restriction readiness follows the same reasoning. □

The set of atomic rules of level n or below is denoted as $\mathbb{S}_n$. Let $\mathfrak{S}_n$ be the full power-set of all rules of level n or lower, $\mathcal{P}(\mathbb{S}_n)$, and let $\mathfrak{S}_\infty$ be the full power-set of all rules, $\mathcal{P}(\mathbb{S})$.

Lemma 4 *For all n, $\mathfrak{S}_n$ is closure ready and restriction ready and so is $\mathfrak{S}_\infty$.*

Proof. Let $S', S^{cl} \in \mathfrak{S}_n^{cl}$ and let S' extend S^{cl}. Note that $S \subseteq S' \cap \mathbb{S}_n \in \mathfrak{S}_n$. We therefore need to show that $S' = (S' \cap \mathbb{S}_n)^{cl}$. Let $R \in (S' \cap \mathbb{S}_n)^{cl}$. It follows that there is a witness for R in S' and because S' is already closed, by Lemma 1 it follows that $R \in S'$. Now let $R \in S'$. We know that either $R \in \mathbb{S}_n$ (in which case we are done) or, because S' is the closure of some set in $\mathfrak{S}_n$, there is a witness for R in $S' \cap \mathbb{S}_n$, but then $R \in (S' \cap \mathbb{S}_n)^{cl}$. The argument for restriction closure is similar. For $\mathfrak{S}_\infty$, simply remove references to the level of $\mathbb{S}$. □

4 Decidability

In this section, we demonstrate that the consequence relation for certain proof-theoretic systems is decidable. Our attention will be focused on the following set of proof-theoretic systems:

Definition 9 *Let DEC be defined as follows:*

1. $\mathfrak{S}_n \in DEC$ for $n \in \mathbb{N}$ and $\mathfrak{S}_\infty \in DEC$.

2. If $\mathfrak{S}$ and $\mathfrak{S}'$ are in DEC then $\mathfrak{S} \cup \mathfrak{S}' \in DEC$.

3. If $\mathfrak{S} \in DEC$ and Φ is a decidable set of atomic rules, then $\{S \in \mathfrak{S} \mid \Phi \subseteq S\} \in DEC$.

By the results of the last section, we know that all proof-theoretic systems in DEC are closure and restriction ready.

Lemma 5 *If $\mathfrak{S} \in DEC$, then $\mathfrak{S}$ is closure and restriction ready.*

Proof. Induction on the definition of DEC. Base case is Lemma 4 and the two induction cases are Lemmas 2 and 3. □

Now recall S^{cl} from Definition 5 and $S^{\{p_i \mid i \in I\}}$ for atomic formulas $\{p_i \mid i \in I\}$ from Definition 7.

Lemma 6 *If $S \in \mathfrak{S} \in DEC$ and $\{p_i \mid i \in I\}$ is the set of all atomic formulas in φ, then $\mathfrak{S}, S \vDash \varphi \Leftrightarrow (\mathfrak{S}^{cl})^{\{p_i \mid i \in I\}}, (S^{cl})^{\{p_i \mid i \in I\}} \vDash \varphi$.*

Proof. By Theorem 2, Theorem 3, and Lemma 5. □

Now $(\mathfrak{S}^{cl})^{\{p_i \mid i \in I\}}$ is not a finite set of finite sets. Our goal is to show that all $\mathfrak{S} \in DEC$ have the finite model property. To achieve this, we need to find a way of reducing $(\mathfrak{S}^{cl})^{\{p_i \mid i \in I\}}$ to a finite set of finite sets. To illustrate how one can find such a structure, consider a sub-case where there is an n such that $\mathfrak{S} \subseteq \mathcal{P}(\mathbb{S}_n)$. In this situation, $\{S \cap \mathbb{S}_n \mid S \in (\mathfrak{S}^{cl})^{\{p_i \mid i \in I\}}\}$ is a finite set of finite sets. This is because for n atomic formulas and level-m rules, there is a function $f(m, n)$ that places a finite bound on the number of rules of that level and number of distinct atomic formulas (if premises are treated as a set and therefore cannot be repeated). For example, for level-1 rules and n atomic formulas, there are $(n \times \sum_{r \leq n} \frac{n!}{r!(n-r)!}) + n$ rules. Note further that because $\mathfrak{S} \subseteq \mathcal{P}(\mathbb{S}_n)$, any rules that occur in an $S \in (\mathfrak{S}^{cl})^{\{p_i \mid i \in I\}}$ and do not occur in any $S' \in \{S \cap \mathbb{S}_n \mid S \in (\mathfrak{S}^{cl})^{\{p_i \mid i \in I\}}\}$ must have been added by closure. Removing them will therefore not change what is provable. Consequently, it follows that $\mathfrak{S} \vDash \varphi \Leftrightarrow \{S \cap \mathbb{S}_n \mid S \in (\mathfrak{S}^{cl})^{\{p_i \mid i \in I\}}\} \vDash \varphi$, where $\{p_i \mid i \in I\}$ denotes the set of all atomic letters in φ.

To generalise this strategy we need to find some set $\mathbb{S}^{DEC_{\{p_i \mid i \in I\}}}$ such that for any $\mathfrak{S} \in DEC$ and φ containing only the atomic letters in $\{p_i \mid i \in I\}$, $\{S \cap \mathbb{S}^{DEC_{\{p_i \mid i \in I\}}} \mid S \in (\mathfrak{S}^{cl})^{\{p_i \mid i \in I\}}\}$ is finite and $\mathfrak{S} \vDash \varphi \Leftrightarrow \{S \cap \mathbb{S}^{DEC_{\{p_i \mid i \in I\}}} \mid S \in (\mathfrak{S}^{cl})^{\{p_i \mid i \in I\}}\} \vDash \varphi$.

To do this we need to introduce two important results from the literature. Together, these results show that, fortunately, if we only care about a finite number of atomic formulas, there will be finitely many non-equivalent rules.

First, note that the *diagram* of the disjunction-free fragment of intuitionistic logic with n propositional variables is the set of equivalence classes of provably equivalent formulas partially ordered by derivability.

Proposition 1 (Urquhart 1974, McKay 1968) *The diagram of the disjunction-free fragment of intuitionistic logic with n propositional variables is finite and decidable.*

For every equivalence class $[\varphi]_\equiv$, one can find a *characteristic* formula. Readers interested in how these are generated are directed to de Lavalette, Hendriks, and de Jongh (2012). All that matters for our purposes is that the characteristic formula of $[p]_\equiv$ is p. The next result allows us to transform the information we now have about intuitionistic logic into information about atomic rules.

Proposition 2 (Piecha et al. 2015) *For every disjunction-free φ, there is an $R^\varphi \in \mathbb{S}$ such that $\mathfrak{S}_\infty, S \vDash \varphi$ if and only if $R^\varphi \in S^{cl}$. And for every $R \in \mathbb{S}$, there is a disjunction-free φ^R such that $\mathfrak{S}_\infty, S \vDash \varphi^R$ iff $R \in S^{cl}$.*

Let $\mathbb{S}_\equiv^{\{p_i \mid i \in I\}}$ be the set of rules generated by the characteristic formulas for the disjunction-free fragment of intuitionistic logic with atomic formulas in $\{p_i \mid i \in I\}$. We then get the following result.

Corollary 2 *For a finite set of atomic formulas $\{p_i \mid i \in I\}$, the set $\mathbb{S}_\equiv^{\{p_i \mid i \in I\}}$ is finite and decidable.*

So $\mathbb{S}_\equiv^{\{p_i \mid i \in I\}}$ is the set we were looking for. It follows that $\mathcal{P}(\mathbb{S}_\equiv^{\{p_i \mid i \in I\}})$ is a finite set of finite sets. To use this result, we need one more fact:

Proposition 3 (Piecha et al. 2015) *Proof-theoretic semantics is sound for intuitionistic logic.*

We also need to know that if two rules are equivalent, they are both either in or out of the closure of an atomic system:

Lemma 7 *Let S be an atomic system, $R \in S$ be an atomic rule and intuitionistic logic prove $\varphi^R \leftrightarrow \varphi^{R'}$, then $R' \in S^{cl}$.*

Proof. Assume $R \in S$. Then by Proposition 2, $\mathfrak{S}_\infty, S \vDash \varphi^R$ and by Proposition 3, $\mathfrak{S}_\infty, S \vDash \varphi^R \leftrightarrow \varphi^{R'}$. So $\mathfrak{S}_\infty, S \vDash \varphi^{R'}$. So by Proposition 2, means that $R' \in S^{cl}$. □

With all this in place, we can now prove our result:

Theorem 4 *If $\mathfrak{S} \in DEC$ then $\{S \cap \mathbb{S}_{\equiv}^{\{p_i|i\in I\}} \mid S \in (\mathfrak{S}^{cl})^{\{p_i|i\in I\}}\}$ is a decidable finite set of finite sets and if $\{p_i \mid i \in I\}$ is the set of atomic formulas in φ, then $\mathfrak{S} \vDash \varphi \Leftrightarrow \{S \cap \mathbb{S}_{\equiv}^{\{p_i|i\in I\}} \mid S \in (\mathfrak{S}^{cl})^{\{p_i|i\in I\}}\} \vDash \varphi$.*

Proof. That $\{S \cap \mathbb{S}_{\equiv}^{\{p_i|i\in I\}} \mid S \in (\mathfrak{S}^{cl})^{\{p_i|i\in I\}}\}$ is a decidable finite set of finite sets is a simple induction on the definition of DEC using Corollary 2.

The proof proceeds by induction on φ. For the base case note that if $(\mathfrak{S}^{cl})^{\{p_i|i\in I\}}, S \vDash p$ then $/p \in (\mathfrak{S}^{cl})^{\{p_i|i\in I\}}$, and it follows that $/p \in \{S \cap \mathbb{S}_{\equiv}^{\{p_i|i\in I\}} \mid S \in (\mathfrak{S}^{cl})^{\{p_i|i\in I\}}\}$.

Now let $(\mathfrak{S}^{cl})^{\{p_i|i\in I\}}, S \vDash \varphi \to \psi$ (which is sufficient by Lemma 6) and $S' \in \{S \cap \mathbb{S}_{\equiv}^{\{p_i|i\in I\}} \mid S \in (\mathfrak{S}^{cl})^{\{p_i|i\in I\}}\}$ be such that it extends $S \cap \mathbb{S}_{\equiv}^{\{p_i|i\in I\}++}$ and $\{S \cap \mathbb{S}_{\equiv}^{\{p_i|i\in I\}} \mid S \in (\mathfrak{S}^{cl})^{\{p_i|i\in I\}}\}, S' \vDash \varphi$. What we need to show is that there is an $S'' \in (\mathfrak{S}^{cl})^{\{p_i|i\in I\}}$ which extends S and $S'' \cap \mathbb{S}_{\equiv}^{\{p_i|i\in I\}} = S'$. We know that there is an S^* in $(\mathfrak{S}^{cl})^{\{p_i|i\in I\}}$ such that $S^* \cap \mathbb{S}_{\equiv}^{\{p_i|i\in I\}} = S'$. We must show S^* extends S. Let $R \in S$. Then R is equivalent to some $R^E \in S'$ but because S^* is closed, it follows that $R \in S^*$ by Lemma 7. □

With this, we have found a way to generate, for any $\mathfrak{S} \in DEC$ and formula φ, a proof-theoretic system that is finite and agrees with $\mathfrak{S}$ on φ. Included in this result are the systems $\mathfrak{S}_1$ and $\mathfrak{S}_2$ which are most likely to be the extensions of atomic rules that Prawitz considered.

References

de Lavalette, G. R., Hendriks, A., & de Jongh, D. H. J. (2012). Intuitionistic implication without disjunction. *Journal of Logic and Computation*, 22(3), 375–404.

McKay, C. G. (1968). The decidability of certain intermediate propositional logics. *Journal of Symbolic Logic*, *33*(2), 258–264.

Piecha, T., de Campos Sanz, W., & Schroeder-Heister, P. (2015). Failure of completeness in proof-theoretic semantics. *Journal of Philosophical Logic*, *44*(3), 321–335.

Prawitz, D. (1971). Ideas and results in proof theory. In J. E. Fenstad (Ed.), *Studies in Logic and the Foundations of Mathematics* (Vol. 63, pp. 235–307). Amsterdam: Elsevier.

Prawitz, D. (1973). Towards a foundation of a general proof theory. In P. Suppes, L. Henkin, A. Joja, & G. C. Moisil (Eds.), *Studies in Logic and the Foundations of Mathematics* (Vol. 74, pp. 225–250). Amsterdam: Elsevier.

Prawitz, D. (1974). On the idea of a general proof theory. *Synthese*, *27*(1/2), 63–77.

Schroeder-Heister, P. (1984). A natural extension of natural deduction. *Journal of Symbolic Logic*, *49*(4), 1284–1300.

Stafford, W. (2021). Proof-theoretic semantics and inquisitive logic. *Journal of Philosophical Logic*, *50*(5), 1199–1229.

Urquhart, A. (1974). Implicational formulas in intuitionistic logic. *Journal of Symbolic Logic*, *39*(4), 661–664.

Will Stafford
Czech Academy of Sciences, Institute of Philosophy
The Czech Republic
E-mail: stafford@flu.cas.cz

Deontic Modal Expressivism: Proof-Theoretic and Model-Theoretic Views

PRESTON STOVALL[1]

Abstract: This article compares two recent expressivist proposals for the deontic modalities: one advanced in proof theory, and the other in model theory. Questions of adequacy are raised for the proof-theoretic proposal, which either do not arise or can be answered by the model-theoretic proposal. At the end of the article, it is proposed that proof theory and model theory may offer complementary, rather than competing, accounts of meaning – the former concerned with intralinguistic relations, and the latter with language-world relations, analogous to the pre-Carnapian distinction between connotation and denotation.

Keywords: deontic modal expressivism, proof theory, model theory

1 Introduction

Although there is a rich history of different approaches to the study of deontic logic (see Hilpinen & McNamara, 2013 for an overview), a widespread position in contemporary philosophy and linguistics understands deontic claims about what is right and wrong, or obliged and forbidden, in terms of what is true at certain possible worlds. Owing largely to the influence of the work of Angelika Kratzer (see, e.g., Kratzer, 1977, 1981, 1991), this tradition is one branch of a more general project, associated with figures like Rudolph Carnap, David Kaplan, David Lewis, Richard Montague, and Barbara Partee at UCLA in the late 1960s and early 1970s, to use model-theoretically characterized possible worlds as a basis for understanding the meanings of different modal or intensional vocabularies.

In contrast to this representational position on deontic claims, which understands them in terms of their (supposed) truth conditions, non-representational approaches to deontic language interpret such claims in terms of

---

[1] My thanks to comments from Luca Incurvati and an anonymous reviewer.

expressions of attitude or commitment. While expressivist approaches to modality risk losing the simplicity and familiarity that truth-conditional possible-world analyses afford, they hold out the hope of offering philosophers and linguists a more realistic model for thinking about the kinds of cognition involved in meaningfully talking about what is and what ought to be or be done. For a claim like "one is obliged to be polite to one's neighbors" does not so much represent the space of possible worlds as express something about the speaker's views on practical rationality.

In the last few years, two lines of research on expressivist interpretations for deontic language have been advanced in some detail, one in proof theory and the other in model theory. According to the semantics I provide in (Stovall, 2021) and (Stovall, 2022a), and more systematically in (Stovall, 2022b):

> "one is obliged to A in C" expresses commitment to a plan of action, understood model-theoretically.

According to the proposal advanced by Incurvati and Schlöder (2017, 2019, 2021),

> "it is right to A in C" expresses commitment to an attitude of approval, understood proof-theoretically.

This raises questions about their relationship, and possible interface. Does one proposal fare better than the other as an analysis of the meaning of deontic claims? Are they compatible, or mutually exclusive?

Incurvati and Schlöder's proof-theoretic approach builds off of (Smiley, 1996), (Rumfitt, 2000), and (Restall, 2005). I give the details in the next section, but in brief, they start with a bilateral natural deduction system employing introduction and elimination rules for speech acts of assertion and linguistic rejection, which count as expressing attitudes of assent and dissent. They use the term "rejection" rather than "linguistic rejection", but part of my aim in this essay is to distinguish different acts of rejection, and I distinguish the proof-theoretic and linguistic notion of rejection bilateralists employ from a model-theoretic and choice-directed alternative introduced below, and discussed in part 3. Against this bilateralist background, where the notion of negation is understood in terms of an attitude of dissent expressed with a speech act of linguistic rejection, Incurvati and Schlöder generalize to a *tri*lateral semantics by adding an attitude of weak assent, expressed by speech acts of weak assertion, so as to give introduction and elimination rules for epistemic modals. This illustrates their method: the metalinguistic resources of different *attitudes* are used, within a proof-theoretic framework employing different classes of *speech act*, to interpret the *semantic functions* of object-language vocabulary. This is accomplished by specifying rules of inference for inferring to and from different claims and the attitudes they express. In (Incurvati & Schlöder, 2021), attitudes of disapproval and approval are used to supply rules for speech acts using deontic vocabulary.

This proof-theoretic expressivist analysis of modality is a promising area of research. But the need to multiply attitudes (from assent and dissent to these plus weak assent, approval, and disapproval) in order to account for different modalities (epistemic, and strong positive and negative deontic forces) comes at the cost of a loss of explanatory lucidity. This can be seen in the analysis Incurvati and Schlöder provide for the strong positive and negative deontic forces: the claim "it is wrong to A in C" expresses disapproval of Aing in C, while the claim "it is right to A in C" expresses approval of Aing in C. These are treated as distinct attitudes, and a principle must be stipulated in order to ensure they are incompatible with one another (details are given below). In addition, as Incurvati and Schlöder point out, their analysis of deontic vocabulary is indirect (Incurvati & Schlöder, 2019, p. 747 and Incurvati & Schlöder, 2021, Section 3.1): rather than expressing an action-guiding attitude directly, as most expressivist analyses of deontic modality do, their analysis allows one to *infer* that such an attitude is expressed. While none of these issues scuttle the program, they do raise considerations worth addressing. They also offer productive points of comparison with model-theoretic deontic modal expressivism.

After presenting Incurvati and Schlöder's proof-theoretic and inferential analysis of deontic vocabulary in part 2, in part 3 I present my model-theoretic and plan-conditional analysis. This latter analysis also uses a notion of rejection, now understood as a volitional act one adopts toward choices, rather than a speech act concerning a propositional content. Call this *agentive rejection*. As I show in part 4, this model-theoretic deontic expressivism is a promising counterpoint to Incurvati and Schlöder's proof-theoretic deontic expressivism: it can explain that in virtue of which the strong positive and strong negative deontic forces are incompatible, for both of these forces (and the weak deontic force for permission) can be explained in terms of the volitional act of agentive rejection; it can account for the semantic function of the intentional modal operator *shall* in its individual and collective modes; and it provides a direct analysis of deontic modality as the expression of our practical rationality. While this might seem to count in favor of model-theoretic modal expressivism, at the end of the essay I suggest that the two views may in fact be not only compatible, but complementary, and I sketch some of the formal terrain that would have to be mapped to determine whether this conjecture is correct.

A few words on terminology and method. In the model-theoretic semantics of part 3, I use the term "agentive rejection" to label a primitive *practical* act that, like bilateralist accounts of negation employing the act of linguistic rejection, is to be understood in terms of a corresponding attitude. But the attitude associated with agentive rejection is not the attitude of dissent that appears in bilateral semantics. Unlike linguistic rejection, which is a speech act, the act of *agentive* rejection is directed at one's choices. The underlying attitudes associated with such acts are neither directed at extralinguistic reality in the word-to-fit-world intentionality characteristic of representation and modeled with possible worlds, nor at intralinguistic content of the sort bilateralist accounts of linguistic rejection are interested in, but rather at one's choices in the world-to-fit-word direction of intentionality (standing here in the shadow of the reflex arc, we begin to see how proof-theoretic and model-theoretic modal expressivism might, like reflective and volitional cognition, be complementary).

Consequently, while I present Incurvati and Schlöder's proof-theoretic interpretation of negation in bilateralist terms (as they do), for ease of exposition in presenting my model-theoretic account I avoid bilateralism and use a single speech act of assertion, which can take both representational and deontic contents. On this view, ordinary descriptive assertions give voice to our representational cognition, modeled by possible worlds and conveyed with ordinary descriptive claims, and the assertion of a sentence governed by a deontic (or intentional) modal operator gives expression to an action-guiding mental state modeled by plans of action. A bilateral semantics could be joined to this approach, but it proves simpler to stick with a single speech act of assertion and treat negation as a matter of content. Given that my focus lies on interpretations for deontic vocabulary, rather than negation in particular, I take this to be an acceptable simplification.[2]

Finally, while I present my view in terms of modal operators for what is obliged, forbidden, and permitted, whereas Incurvati and Schlöder offer analyses for predicates concerning what is right, wrong, and tolerated, I do

---

[2] See the first appendix of (Stovall, 2022b) for the resolution of problems often thought to plague plan-conditional expressivist analyses of negation.

not put any significance on this difference, at least at the level of detail that concerns me here.

2 Proof-theoretic deontic modal expressivism: Inferential expressivism

Incurvati and Schlöder's account takes off from the bilateralist's key idea, that the speech act of *asserting* ϕ expresses an attitude of *assent* to ϕ, while the speech act of *rejecting* ϕ expresses an attitude of *dissent* from ϕ (my reconstruction follows the account given in their 2021). This allows for the following account of negation:

> **Bilateralist Negation:** The speech act of *asserting* $\neg\phi$ expresses an attitude of *dissent* from ϕ.

This account is spelled out proof-theoretically as follows. Where $+$ marks an attitude of assent expressed with a speech act of assertion, and $\ominus$ marks an attitude of dissent expressed with a speech act of linguistic rejection, the assertion of $\neg\phi$ can be understood as an expression of dissent from ϕ, in the linguistic rejection of ϕ; just so, the linguistic rejection of ϕ can be understood as an expression of dissent from ϕ, in the assertion of $\neg\phi$:

$$\frac{+\neg\phi}{\ominus\phi}\,(+\neg E) \qquad \frac{\ominus\phi}{+\neg\phi}\,(+\neg I)$$

Correlative elimination and introduction rules are given for the linguistic rejection of negation, as well as rules for the assertion and linguistic rejection of other logical operators. And because assertion and linguistic rejection are two distinct and incompatible speech acts, Smileyan coordination principles are needed to relate them:

$$\frac{+\phi \quad \ominus\phi}{\bot}\,Rejection \qquad \frac{[+\phi]}{\vdots} \\ \frac{\bot}{\ominus\phi}\,SR_1 \qquad \frac{[\ominus\phi]}{\vdots} \\ \frac{\bot}{+\phi}\,SR_2$$

Moving from a bilateral to a trilateral semantics, rules are then given for interpreting the weak epistemic modal operator *perhaps*, symbolized by $\Diamond$, as the expression of *weak assent*, symbolized by $\oplus$:

$$\frac{\oplus\phi}{+\Diamond\phi}\ (+\Diamond I) \qquad \frac{+\Diamond\phi}{\oplus\phi}\ (+\Diamond E)$$

$$\frac{\oplus\phi}{\oplus\Diamond\phi}\ (\oplus\Diamond I) \qquad \frac{\oplus\Diamond\phi}{\oplus\phi}\ (\oplus\Diamond E)$$

Finally, because the underlying semantics now makes use of three distinct attitudes, rules for relating weak assent to assent, dissent, and the logical operators are then given.

Incurvati and Schlöder argue that the resulting system satisifies a number of desiderata for a semantics employing these notions, and they give rules for interpreting the predicates "right" and "wrong" in terms of two further attitudes: approval and disapproval. I stick to their terminology in the presentation, but for the purposes of exposition I take the predicates "right" and "wrong" to be interchangeable with modal operators "it is obliged that" and "it is forbidden that" – what is important is that the strong positive and negative deontic forces be distinguished from each other, and from the weak force of permsission or what is merely okay (neither right nor wrong – their term is "tolerated"). Where $\mathcal{A}$ marks the attitude of *approval*, $\mathcal{D}$ marks the attitude of *disapproval*, and A ranges over actions:

$$\frac{\mathcal{A}A}{+rightA}\ (+rightI) \qquad \frac{+rightA}{\mathcal{A}A}\ (+rightE)$$

$$\frac{\mathcal{D}A}{+wrongA}\ (+wrongI) \qquad \frac{+wrongA}{\mathcal{D}A}\ (+wrongE)$$

In order to establish the inconsistency of asserting that something is both right and wrong, they appeal to the following axiom:

RW-Contradiction: $+\neg(rightA\ \&\ wrongA)$

Finally, what is merely okay (tolerated, or permitted but not obliged) is defined in terms of what is not wrong: to assert that something is not wrong is to express dissent from the claim that it is wrong.[3] As they point out, there

---

[3] Strictly speaking, their discussion of asserting that A is not wrong proceeds by way of a notion of *stable* dissent, defined in terms of ordinary dissent ($\ominus$) from the epistemic possibility ($\Diamond$) that A is wrong:

it is not the case that A is wrong $=_{def.}\ \ominus\Diamond wrongA$

I suppress this permutation here.

is thus no need for a third attitude to make sense of the weak deontic force of permission or tolerance.

Inferential expressivism offers a sophisticated analysis for a range of logical vocabularies in terms of corresponding attitudes. But a number of bridge principles are needed to make sense of the interaction among these different attitudes, as well as the principle of RW-Contradiction to ensure the inconsistency of the thought that something is both right and wrong, and this complicates the semantics. Also, as noted above, the relationship between the assertion of a deontic claim and the attitude expressed here is *indirect*, mediated by an inference from the claim to the conclusion that the attitude is expressed. But when I make a claim like "bringing an umbrella is the right thing to do", or "one is obliged to be polite to one's neighbors", I seem to be directly giving voice to something about my agency, and my ability to consider what to do were I other people.

Additional questions are raised over how the expression of approval and disapproval explains the meaning of the moral "right" and "wrong". For why should approval and disapproval themselves align with what one regards as right and wrong? After all, it does not seem that the attitude of approval by itself licenses an inference to a claim that something is right, or an attitude of disapproval license an inference to a claim that it is wrong, as the rules $+rightI$ and $+wrongI$ require. Surely one can approve (disapprove) of something without committing oneself to the claim that it is right (wrong) or ought (ought not) be done; and it would appear that one could approve of both doing something and not doing it, whereas it's not clear that one could hold that one ought to both do something and not do it.[4] And one might like to know more about the attitudes inferential expressivism posits for interpreting deontic modality, for the view requires two distinct attitudes and an axiom of RW-Contradiction to relate them to one another.

In part 3, I lay out my model-theoretic deontic expressivism and consider how it fares against these questions in parts 4 and 5.

3 Model-theoretic deontic modal expressivism: Deontic-intentional hyperstate semantics

Definition 1 *A possible world $w \in W$ is a maximally determinate state of affairs.*

---

[4]My thanks to a reviewer for raising this concern.

Definition 2 *An intentional hyperplan $h_I \in H_I$ is a maximally determinate plan of action such that for every agent α, every circumstance C, and every action A open to α at C, α either (exclusively)*

- *chooses to A in C, or*
- *chooses not to A in C*

Unlike intentional hyperplans, the choices that define deontic hyperplans are distinguished in terms of two choice attitudes: single-mindedness and indifference. To single-mindedly choose to A is to *reject*, in the agentive sense of the term, every choice incompatible with A.

Definition 3 *An agent α single-mindedly chooses to A in C just in case α agentively rejects every choice B incompatible with A.*

Single-mindedness is an attitude one can have in acting, analogous to the attitudes of assent and dissent that speech acts of assertion and linguistic rejection give voice to. Agentive rejection is then analogous to the speech act of linguistic rejection, it being an act directed at one's choices rather than at the contents of speech or thought. While I take agentive rejection as a primitive notion, using it to define single-mindedness and indifference (see definition 4), it can be glossed as an act undertaken with a negative affective valence directed at various choices. To single-mindedly choose to do something, then, is to exercise agentive rejection directed at every choice incompatible with what one does. It follows that the act of choosing under the attitude of single-mindedness is an act of self-government: one binds oneself to a course of action by refusing to allow oneself to do anything incompatible with it.

For example, the single-minded choice to be in Prague by 8 a.m. proceeds by my binding myself to that course of action in suppressing inclination to, among other things, sleep in. This binding or self-government is brought off by the act of agentively rejecting every choice incompatible with what one has set oneself to do. The idea that human beings are capable of such maximally determinate mental states is an idealization, of course, and there is no pretense that anyone other than the angels – and perhaps Kant – has exercized what the model posits as perfectly single-minded practical cognition. Still, this is an idealization that (we may profitably suppose) is approximately realized in the higher-order executive functioning and self-control characteristic of human cognition.[5]

Indifference can be defined in terms of single-mindedness: fix all of an agent's single-minded choices, and she chooses indifferently to A at a point of choice just in case there is some action B that is both incompatible with A, and which she could have undertaken without changing any of her single-minded choices. In this case, the choice between A and B is indifferent, and the agent chooses at once to *reject* rejecting each of them – in the sense of agentive rejection (this should be taken as implied where needed in what follows).

Definition 4 *An agent α capable of single-mindedness indifferently chooses to A in C just in case there is some choice B incompatible with A that α could have made without changing any of her single-minded choices. In this case, the agent agentively rejects agentively rejecting both A and B.*[6]

Notice that even when one chooses to A indifferently, that choice and the indifferent choice not to B are alike instances of rejecting rejecting the choice in question. So when I have decided single-mindedly to be in Prague by 8 a.m., but have decided that it is indifferent to me whether I take the bus or the train, then my choice between these two options is one of both rejecting rejecting taking the bus, and rejecting rejecting taking the train. Because rejection iterates in this way, and because it is treated as a negative valence, it can be interpreted as a complementarity operator so as to deliver the required entailments for a deontic modal logic (see below for some illustrations, and Stovall, 2022a for the details). Deontic modal claims can then be understood as expressions of the attitudes associated with this notion of agentive rejection. To see this, we first have to introduce the notion of a deontic hyperplan, and then specify the semantic interpretants employed by the models.

Definition 5 *A deontic hyerplan $h_D \in H_D$ is a maximally determinate plan of action such that for every agent α, every circumstance C, and every action A open to α at C, α either (exclusively)*

---

[5] There is much more to be said about the identification and individuation of choices on deontic hyperplans. See chapters 4 and 5 of (Stovall, 2022b) for a more thorough discussion, and sections 3 and 6 of (Stovall, 2022a) for some highlights.

[6] The restriction to agents capable of single-mindedness is to ensure that the merely intentional animal does not trivially count as acting under an attitude of indifference.

- *single-mindedly chooses to A in C, or*
- *single-mindedly chooses not to A in C, or*
- *indifferently chooses to A in C, or*
- *indifferently chooses not to A in C*

Definition 6 *A deontic-intentional hyperstate $\langle w, h_D, h_I \rangle$ is an ordered triple of a possible world, a deontic hyperplan, and an intentional hyperplan.*

Atomic descriptive, pure intentional, and pure deontic claims can be evaluated in terms of deontic-intentional hyperstates (I sometimes speak simply in terms of hyperstates). Intuitively, descriptive claims represent the world in the word-to-fit-world direction of intentionality, while intentional and deontic claims give expression to the world-to-fit-word intentionality of practical cognition, where *deontic* practical cognition involves an exercise of single-mindedness and agentive rejection. The claim that one is obliged to A in C, for instance, gives expression to the practical attitude of rejection adopted toward every choice incompatible with doing A in C, and modeled by a plan to choose single-mindedly to A in C from the perspective of everyone denoted by "one". Just so, the claim that one is forbidden to A in C expresses rejecting choosing to A in C, modeled by the plan to single-mindedly choose not to A in C. Unlike the speech act of rejection used in Incurvati and Schlöder's proof theory, this model-theoretic analysis of agentive rejection iterates in a way that naturally gives rise to the distinction between the strong and weak deontic forces. For to think that one is permitted to A in C is to *reject rejecting* doing A in C, modeled by the set of plans where one either single-mindedly chooses to A in C, or *indifferently* chooses whether or not to A in C.

More precisely, where ι is a metalingusitic variable ranging over the singular and plural first-person pronouns, where the modalities are read *de dicto*, and where "it is obliged that", etc. connotes the endorsement of a norm and not merely a report that a norm is in place, we have the following:

Definition 7

$[\![d]\!] =_{def.} \{\langle w, h_D, h_I \rangle : d \text{ is true at } w\}$

$[\![\iota \text{ shall } A \text{ in } C]\!] =_{def.} \{\langle w, h_D, h_I \rangle : \iota \text{ chooses to } A \text{ in } C \text{ on } h_I\}$

$[\![\text{it is obliged that everyone } A \text{ in } C]\!] =_{def.} \{\langle w, h_D, h_I \rangle : \text{every } \alpha \text{ single-mindedly chooses to } A \text{ in } C \text{ on } h_D\}$

$[\![\text{it is forbidden that everyone A in C}]\!] =_{def.} \{\langle w, h_D, h_I\rangle: \text{some } \alpha \text{ single-mindedly chooses not to A in C on } h_D\}$[7]

$[\![\text{it is permitted that everyone A in C}]\!] =_{def.} \{\langle w, h_D, h_I\rangle: \text{every } \alpha \text{ either single-mindedly chooses to A in C on } h_D, \text{ or indifferently chooses whether or not to A in C on } h_D\}$

An example may help illustrate the analysis for the deontic fragment. The claim that it is obliged of conference attendees that they arrive on time expresses commitment to a plan to choose single-mindedly to arrive on time at any context in which one is a conference attendee. Commitment to this transperspectival plan is marked with an act of agentively rejecting everything incompatible with attending on time. The additional claim that conference attendees are permitted to sit in either the front or the back expresses commitment to a plan to choose indifferently where to sit, by rejecting rejecting each choice, in the sense that either choice could be undertaken without breaking the single-minded commitment to arrive on time. Even where one has a preference or inclination to sit in one place, the choice will be indifferent in the stipulated sense, insofar as one could have taken the other option without changing one's single-minded commitment to arrive on time.

Introduction of the Boolean operators is straightforward, resulting in a recursively computable compositional semantics that distinguishes the contents of representational, intentional, and deontic cognition in both atomic and logically complex sentences.

Definition 8

$[\![\neg \phi]\!] =_{def.} [\![\phi]\!]^C$

$[\![\phi \wedge \psi]\!] =_{def.} [\![\phi]\!] \cap [\![\psi]\!]$

$[\![\phi \vee \psi]\!] =_{def.} [\![\phi]\!] \cup [\![\psi]\!]$

$[\![\phi \rightarrow \psi]\!] =_{def.} [\![\phi]\!]^C \cup [\![\psi]\!]$

Furthermore, it is easily established that negation functions as a complement operator on both representational and practical contents, that the strong and weak deontic modalities are duals of one another with respect to negation, and

---

[7] Where "forbidden" means "obliged to not", where the universal and existential quantifiers are given the usual definitions, and where double negation holds, this formulation with an existential quantifier in the definiens is correct. See Appendix 3 of (Stovall, 2022b).

that something is permitted if it is obliged (see Stovall, 2022a and Appendix 3 of Stovall, 2022b). Here is a proof sketch of the following principle:

> it is obliged that everyone A in C *iff* it is not the case that it is permitted that not everyone A in C

To show that this holds, we show that one and the same set of hyperstates models the claims on each side of the biconditional. The lefthand claim is modeled by the set of hyperstates whose deontic hyperplans are such that everyone chooses single-mindedly to A in C. What must be established is that this set is identical to the complement, in the space of hyperstates, of the set of hyperstates that models the following:

> it is permitted that not everyone A in C

This is in turn equivalent to:

> it is permitted that someone not A in C

This last claim is modeled by the set of hyperstates whese deontic hyperplans are such that someone either chooses single-mindedly not to A in C, or chooses indifferently whether or not to A in C. The complement of this set of hyperstates is the set whose deontic hyperplans are such that everyone chooses single-mindedly to A in C, which is the required result.

The semantics also obeys a disjunctive normal form theorem, which provides a straightforward analysis of any sentence of arbitrary logical complexity as a disjunction of conjunctions concerning atomic ways the world *is* and *plans* for how to act in those situations. If the underlying logic is quantificational, *de re* and *de dicto* contexts can be marked, and mixed descriptive/practical contents modeled (e.g., "there is someone not in attendance and he or she ought to be here"). These results are established in (Stovall, 2022a).

Finally, it is worth pointing out why this framework does not require an axiom like RW-Contradiction. Recall that this axiom states that it is not the case that Aing is both obliged and forbidden (or right and wrong, in Incurvati and Schlöder's terminology). By taking each of the positive and negative deontic modalities as primitive, understood through distinct attitudes of approval and disapproval, Incurvati and Schlöder have no basis for explaining why something cannot be both obliged and forbidden, and so they must stipulate that RW-Contradiction holds. But because hyperstate deontic modal

Deontic Modal Expressivisim

expressivism treats the deontic modalities in terms of agentive rejection, and the attitudes of single-mindedness and indifference it characterizes, the principle behind this axiom can be derived from that analysis. What must be shown is that there is no hyperstate whose deontic hyperplan has someone both single-mindedly choosing to A in C, and single-mindedly choosing not to A in C. Definition 5 establishes this very thing, on the basis of the fact that one cannot both do and not do something at a single point of choice.

4 Comparing proof-theoretic and model-theoretic deontic expressivisms

This model-theoretic deontic expressivism either offers answers to, or avoids the need to address, the questions raised against proof-theoretic deontic expressivism in part 2, and without needing to multiply attitudes (the epistemic modalities can be given a standard possible-world analysis).

First, the inconsistency of thinking that something is obliged and forbidden (understood as expressions of commitment and not merely reports on the existence of a norm) is accounted for by the prosaic feature of practical agency that one cannot both do and not do something (in the relevant sense) – no stipulated axiom like RW-Contradiction is needed. Second, the direct connection between deontic cognition and agency is accounted for in terms of the fact that an assertion like "one is obliged to be polite to one's neighbors" gives expression to the practical attitude of single-mindedness. That attitude is not merely inferred on the basis of such an assertion; it is voiced with that claim. Additionally, specifically *moral* claims, and the attitudes expressed by them, are naturally accounted for in terms of these plans of action. For a plan is the very sort of thing that guides one's action in the pursuit of some end, and a moral plan involves a kind of self-government characteristic of the moral law: the one who plans single-mindedly from a *moral* point of view is not conditioning that plan on one's instrumental goals.

These concerns do not conclusively establish that model-theoretic plan-conditional hyperstate expressivism about deontic modality is to be preferred over proof-theoretic inferential expressivism. But in one regard at least it does seem that hyperstate semantics has an advantage over inferential expressivism. Whereas inferential expressivism posits two distinct attitudes for the strong positive and negative deontic forces – approval and disapproval – and defines the weak deontic force in terms of dissent from disapproval, hyperstate semantics accounts for the semantic function of the strong positive, strong negative, and weak deontic forces in terms of a single volitional act of *agentive rejection*, and the kinds of attitudes that can be understood in terms of it. To say that one is obliged to A in C is to express commitment to a plan to single-mindedly choose to A in C, which is to reject not choosing to A in C; to say that one is obliged not A in C is to express commitment to a plan to single-mindedly choose not to A in C, which is to reject choosing to A in C; and to say that one is permitted to A in C is to express commitment to a plan to either single-mindedly choose to A in C, or to indifferently choose whether or not to A in C, which is to reject rejecting choosing to A in C. Both perspectives must appeal to a notion of negation, whether accounted for in terms of content or attitude, but agentive rejection allows for a unified treatment of the deontic modal forces.

Unlike bilateralist accounts of negation, there appears to be a genuine gain in simplicity in the model-theoretic approach here that the proof-theoretic approach lacks. In the case of negation, we have a reason to treat assertion and linguistic rejection as distinct acts with the associated attitudes of assent and dissent, for there is a trade-off either way: one either takes a single act and its attitude (assertion and assent), and treats negation as an additional content; or one adds an additional act and its attitude (linguistic rejection and dissent), while then explaining negation in those terms. So in the case of bilateralist accounts of negation, positing distinct attitudes can be motivated on the basis of the fact that one *gains* in simplicity concerning content, by dispensing with a content-based account of negation.

But now consider the case of inferential expressivism and the deontic forces. Here deontic claims are understood in terms of two distinct attitudes, approval and disapproval. RW-Contradiction must then be imposed to ensure these attitudes are jointly inconsistent. By contrast, hyperstate expressivism appeals to two attitudes – single-mindedness and indifference – but these are explained in terms of the act of agentive rejection: one single-mindedly chooses to A in C just in case one rejects everything incompatible with that choice, while one chooses indifferently to A in C just in case one is capable of choosing single-mindedly (that is, capable of agentive rejection), and there is some act B incompatible with A that one could have made without changing any of one's single-minded choices. Unlike the case of inferential expressivism, the two distinct attitudes of hyperstate expressivism admit of a single treatment in terms of the act of agentive rejection. And this act, together with the single-minded and indifferent choice attitudes associated with it, and the fact that one cannot both do something and not do it, explains why it is inconsistent to assert that one is both obliged to and forbidden from doing something.

5 Proof theory and model theory as complementary semantic frameworks

These considerations may make it seem that the model-theoretic framework of deontic-intentional hyperstate semantics is a more promising basis for developing deontic modal expressivism than the proof-theoretic framework of inferential expressivism. But I suspect that model theory and proof theory are complementary resources for thinking about linguistic meaning, and that there is a kind of historically contingent distortion in our thinking to see proof theory and model theory (or inference and reference) as competitor semantic frameworks.

For centuries philosophers and logicians distinguished concept-world relations of meaning from concept-concept relations – e.g. in terms of "extension" and "intension", or "denotation" and "connotation", or "reference" and "sense". Following Carnap's decision to use Leibniz's term "intension" to label functions from state descriptions to extensions in *Meaning and Necessity* (1947), as a replacement for Frege's notion of *Sinn*, the non-denotational dimension of meaning has been almost universally reconstructed within model theory. A full story of the predominance of model-theoretic possible-world intensions would have to include both a discussion of its development in the 1950s and '60s in the work of figures like Barcan Marcus, Kripke, Lewis, and Stalnaker, and its influence on linguistics in the 1970s and '80s with the widespread application of Montague semantics and the work of figures like Kratzer and Partee. The adequacy of such a story would turn on its ability to incorporate an account of the development of proof theory during this period.

I leave this interweaving of model-theoretic and proof-theoretic storylines aside, as the design of the tapestry can be appreciated even in broad outline. For if we make a distinction between the reference or denotation of a sentence, interpreted in terms of relations between *word* and *world* laid down by a model employing possible worlds and plans of action, and the sense or connotation of a sentence interpreted in terms of *word-word* relations laid down by proof theory's rules of inference, then it is open to interpret Incurvati and Schlöder's account as targeting the connotations of deontic claims, while seeing the model-theoretic hyperstate account as targeting their agency-coordinating denotations. This has the appealing consequence of explaining that in virtue of which Incurvati and Schlöder's proof theory offers an indirect analysis of deontic terminology, whereas the hyperstate account is direct: for we can follow Frege and think of connotation or sense (in proof theory) as a mode of determination of denotation or reference (in model theory) – it is because we *approve* of doing A in C, so the thought goes, that we *plan to single-mindedly* do so.

This would not of itself answer the questions raised in part 4 about the adequacy of the proof-theoretic analysis, and in that regard we may have a better grip on the model-theoretic denotations of deontic modality offered by hyperstate semantics. But it *is* to say that the attempt to answer those questions need not be framed in terms that would rule out the hyperstate interpretation of deontic modality. And we may gain more precision in thinking about linguistic meaning – and human cognition – in terms of a theory of meaning employing the resources of both proof theory and model theory, where model-theoretic denotations are further subdivided into those having the word-to-fit-world intentionality of representational cognition (modeled with possible worlds) and those having the world-to-fit-word intentionality of practical cognition (modeled with plans of action).[8] The result is a picture of meaning that, like the reflex arc, involves a reflexive or internal reflective moment, and two moments of world-regarding content.

I hope the foregoing discussion has done enough to situate these two expressivisms in their respective terrains as to both provide a map for, and exhibit the value in, further investigation of their relationship. More generally, the fact that model-theoretic semantics provides a framework for theorizing about the word-to-fit-world and world-to-fit-word intentionalities of representation and agency, while proof theory affords a means for interpreting content in terms of intralinguistic rules of inference, suggests three things:

1. there is interesting work to be done in juxtaposing model-theoretic and proof-theoretic expressivist positions against one another;

2. the pre-Carnapian distinction between denotation and connotation should be reframed with the tools of model theory and proof theory;

3. to do that is to gain both a surer grip on what these notions meant for centuries before the advent of possible-world semantics, and a

---

[8]This is not to say that human cognition itself is ever so bifurcated, but rather that it proves useful to think of human cognition in terms of these idealizations.

more sophisticated understanding of ourselves as sensing, thinking, and acting beings.

References

Carnap, R. (1947). *Meaning and Necessity: A Study in Semantics and Modal Logic*. Chicago: University of Chicago Press.

Hilpinen, R., & McNamara, P. (2013). Deontic logic: A historical survey and introduction. In D. Gabbay, J. Horty, X. Parent, R. van der Meyden, & L. van der Torre (Eds.), *Handbook of Deontic Logic and Normative Systems* (pp. 639–650). London: College Publications.

Incurvati, L., & Schlöder, J. J. (2017). Weak rejection. *Australasian Journal of Philosophy*, *95*(4), 1–20.

Incurvati, L., & Schlöder, J. J. (2019). Weak assertion. *The Philosophical Quarterly*, *69*(277), 741–770.

Incurvati, L., & Schlöder, J. J. (2021). Inferential expressivism and the negation problem. In R. Shafer-Landau (Ed.), *Oxford Studies in Metaethics* (Vol. 16, pp. 80–107). Oxford: Oxford University Press.

Kratzer, A. (1977). What 'must' and 'can' must and can mean. *Linguistics and Philosophy*, *1*(3), 337–355.

Kratzer, A. (1981). The notional category of modality. In H. J. Eikmeyer & H. Rieser (Eds.), *Words, Worlds, and Contexts* (pp. 38–74). Berlin: de Gruyter.

Kratzer, A. (1991). Modality. In A. von Stechow & D. Wunderlich (Eds.), *Semantics: An International Handbook of Contemporary Research* (pp. 639–650). Berlin: de Gruyter.

Restall, G. (2005). Multiple conclusions. In P. Hájek, L. Valdés-Villanueva, & D. Westerståhl (Eds.), *Logic, Methodology and Philosophy of Science: Proceedings of the Twelfth International Conference* (pp. 189–205). London: King's College Publications.

Rumfitt, I. (2000). 'Yes' and 'No'. *Mind*, *109*(436), 781–823.

Smiley, T. (1996). Rejection. *Analysis*, *56*(1), 1–9.

Stovall, P. (2021). The metaphysics of practical rationality: Intentional and deontic cognition. *Journal of the American Philosophical Association*, *7*(4), 549–568.

Stovall, P. (2022a). Modeling descriptive and deontic cognition as two modes of relation between mind and world. *Pacific Philosophical Quarterly*, *103*(1), 156–185.

Stovall, P. (2022b). *The Single-Minded Animal: Shared Intentionality, Normativity, and the Foundations of Discursive Cognition.* New York: Routledge.

Preston Stovall
University of Hradec Králové, Faculty of Philosophy and Social Science
The Czech Republic
E-mail: `preston.stovall@uhk.cz`

www.ingramcontent.com/pod-product-compliance
Lightning Source LLC
Chambersburg PA
CBHW062217080426
42734CB00010B/1922